Denoising - New Insights

Edited by Mourad Talbi

Published in London, United Kingdom

Denoising - New Insights
http://dx.doi.org/10.5772/intechopen.102307
Edited by Mourad Talbi

Contributors
Pavel Němeček, Kenji Hara, Kohei Inoue, Avaragollada Puravarga Mathada Prasanna Kumar, S.M. Vijaya, Mohamed Yahia, Tarig Abdelgayoum Ali, Mourad Talbi, Riadh Baazaoui

First published in London, United Kingdom, 2023 by IntechOpen
IntechOpen is the global imprint of INTECHOPEN LIMITED, registered in England and Wales, registration number: 11086078, 5 Princes Gate Court, London, SW7 2QJ, United Kingdom

British Library Cataloguing-in-Publication Data
A catalogue record for this book is available from the British Library

Additional hard and PDF copies can be obtained from orders@intechopen.com

Denoising - New Insights
Edited by Mourad Talbi
p. cm.
Print ISBN 978-1-83768-029-0
Online ISBN 978-1-83768-030-6
eBook (PDF) ISBN 978-1-83768-031-3

Meet the editor

Mourad Talbi is an Associate Professor of Signal and Image Processing at the Center of Researches and Technologies of Energy of Borj Cedria, Tunis, Tunisia. He obtained a master's degree in Automatics and Signal Processing from the National Engineering School of Tunis (ENIT) in 2004. He obtained a Ph.D. and HDR in Electronics from the Faculty of Sciences of Tunis in 2010 and 2015, respectively. Dr. Talbi is a research member of the Laboratory of Nanomaterials and Systems for Renewable Energies (LaNSER), CRTEn, Tunisia.

Contents

Preface

Signal and image denoising is a branch of data processing. It is an exciting area of research in which numerous signal and image denoising techniques are being developed.

This book presents the state of the art in the field and examines several new denoising techniques. It is divided into two sections. The first section discusses signal denoising and some novel image denoising techniques. The second section deals with image denoising. This volume is a useful resource for engineers and researchers in the fields of signal and image processing.

The editor thanks all the authors for their high-level contributions and their proactive collaboration in the realization of this book.

Mourad Talbi
Research Member,
Laboratory of Nanomaterials and Systems for Renewable Energies (LaNSER),
Tunis, Tunisia

Professor of Signal and Image Processing,
Center of Researches and Technologies of Energy (CRTEn),
Tunis, Tunisia

Section 1

Signal Denoising

Chapter 1

Introductory Chapter: Signal and Image Denoising

Mourad Talbi

1. Introduction

Both signal and image are unfortunately degraded by different factors that affect as noise during acquisition or transmission. Those noisy effects decrease the performance of visual and computerized analysis. It is clear that cancelling the noise from the signal facilitates its processing. The denoising process can be described as to cancel the noise while retaining and not distorting the quality of processed signal or image [1–4]. The conventional manner of denoising for noise cancelling consists in applying a band/low-pass filter with cut-off frequencies. Though conventional filtering methods are capable to suppress a relevant of the noise, they are not able when the noise is located in the band of the signal to be processed. Consequently, numerous denoising techniques were introduced in order to overcome this problem. The algorithms and processing approaches employed for signals can be also used for images and this is due to fact that an image is viewed as a two-dimensional signal. Consequently, the signal processing methods for one-dimensional signals can be adapted for processing two-dimensional images. Due to the fact that the origin and non-stationarity of the noise corrupt the signal, it is not easy to model it. Nevertheless, when the noise can be considered as stationary, an empirically recorded signal that is degraded by an additive noise is formulated as follows [1]:

$$y(j) = x(j) + \sigma \cdot \varepsilon(j), j = 0, 1, \ldots, n - 1 \quad (1)$$

With $y(j)$ is the noisy signal, $x(j)$ is the clean signal and $\varepsilon(j)$ are independently normal random variables and σ designates the level noise corrupting (j). The noise can be modeled as stationary independent zero-mean white Gaussian variables [5, 6]. If this model is employed, the objective of noise cancellation consists in reconstructing $x(j)$ from a finite set of $y(j)$ values without considering a particular structure for the signal. The commonly used approach for noise cancellation models noise as a high frequency signal corrupting in additive manner, the clean signal. These high frequencies can be bringing out employing Fourier transform, ultimately cancelling them by an adequate filtering. This noise cancelling method is conceptually clear and efficient since it is depending only on computing DFT (Discrete Fourier Transform) [7]. However, there is some issue that should be considered. The most important having same frequency since the noise owns important information in the original signal. Filtering out these frequency components introduces noticeable information loss of the desired signal. It is clear that a technique is strongly needed for preserving the prominent part of the signal having relatively high frequencies as the noise has. As an example, the wavelet-based noise removal approaches have provided this prominent part conservation. De-noising of natural images degraded by Gaussian Noise employing wavelet based denoising techniques are very efficient due to the fact that it is able to capture the energy of a signal in few energy transform values. The wavelet de-noising scheme thresholds the wavelet coefficients arising from the standard discrete wavelet transform [8]. In Ref. [8], it was introduced to investigate the suitability of different wavelet bases and the size of different neighborhood on the performance of image denoising techniques in term of peak signal-to-noise ratio (PSNR) [8].

In Ref. [9], Di Liu and Xiyuan Chen introduced an image denoising technique applying an ameliorated bidimensional empirical mode decomposition (BEMD) and using soft interval thresholding. At first step, a noise compressed image is constructed. After that, this noise compressed image is decomposed by applying BEMD into a series of intrinsic mode functions (IMFs), which are separated into signal-dominant IMFs and noise-dominant IMFs employing a similarity measure based on $\ell 2$-norm and a probability density function, and a soft interval thresholding is employed in adaptive manner for cancelling the noise inherent in noise-dominant IMFs. The denoised image is finally obtained *via* the combination of the signal dominant IMFs and the denoised noise dominant IMFs. The performance of this image denoising technique [1] was applied to multiple images with different sorts of noise, and the results obtained from the application of this technique [1] were compared to those obtained from the application the some traditional techniques in different noisy environments. Simulation results in terms of peak signal-to-noise ratio, mean square error, and energy of the first IMF, proved that this denoising technique [9] outperforms the other denoising techniques.

Hybridization of the BEMD with denoising approaches has been introduced in the literature as an efficient image denoising technique.

In Ref. [10], Student's probability density function was proposed in the calculation of the Mean Envelope of the data during the BEMD sifting process for making it robust to values that are far from the mean. The obtained BEMD was named tBEMD. To prove the efficiency of the tBEMD, many image denoising approaches were used in the tBEMD field. Among these approaches, we can mention the discrete wavelet transform (DWT), fourth-order partial differential equation (PDE), linear complex diffusion process (LCDP), and nonlinear complex diffusion process (NLCDP). For experiments, a standard digital image and two biomedical images are considered. The original images are degraded by additive Gaussian Noise with three diverse levels. Based on PSNR (peak signal-to-noise ratio), the obtained results show that DWT, PDE, LCDP, and NLCDP, all perform better in the tBEMD domain compared to the conventional BEMD domain. Moreover, the tBEMD is faster than conventional BEMD in case where the noise level is low. However, in case where it is high, the calculation cost in terms of processing time is similar. The efficiency of the presented approach makes it promising for clinical applications.

This book is intended for engineers and researchers in the fields of signal and image processing. Indeed, this book deal with a large number of signal and image denoising techniques. These techniques include an innovative image denoising approaches.

2. Examples of signal and image denoinsing

In this section, we will give some examples of signal and image denoising obtained from the application of the discrete wavelet transform (DWT).

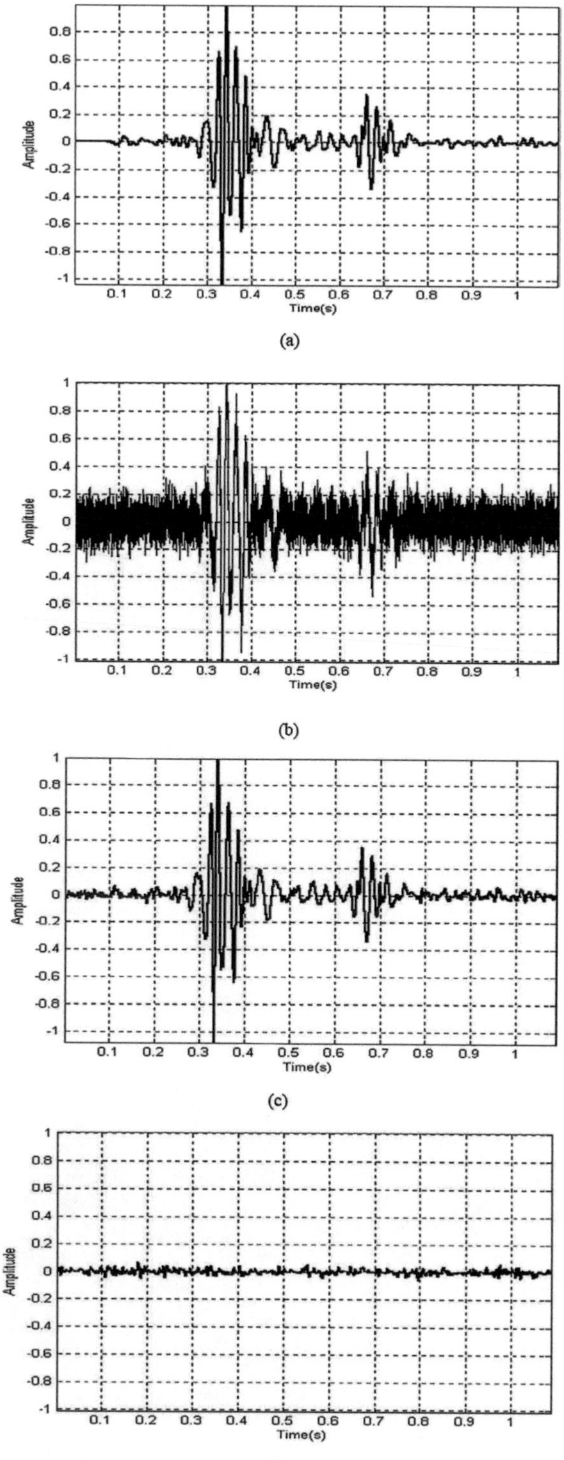

Figure 1.
An example of PCG denoising using DWT [1]: (a) clean PCG signal, (b) noisy PCG signal, (c) denoised PCG signal, (d) difference between the original and the denoised signal.

2.1 Phonocardiogram denoising

The acoustical vibrations records from the heart, acquired through microphones from human chest, named phonocardiogram (PCG), consist of both the murmurs and the heart sounds. Those records of acoustic signals are unfortunately corrupted by diverse factors which effecting as noise. Those effects cause the decreasing of the performance of visual and computerized analysis [1, 11, 12].

Figure 1 illustrates an example of PCG denoising using DWT.

According to **Figure 1**, the noise is considerably reduced and the waveform of the original signal is conserved because the difference between the original and the denoised signals is very small. Consequently, the denoising technique based on thresholding in DWT domain and applied in Ref. [1], shows its performance in noise reduction while conserving the information contained in the original PCG signal.

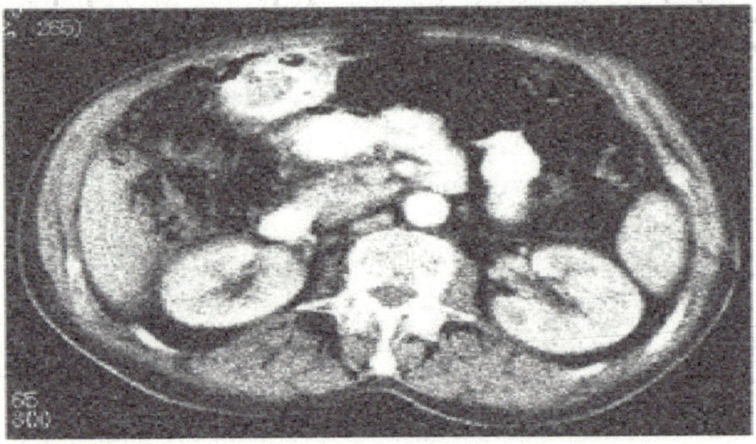

(a)

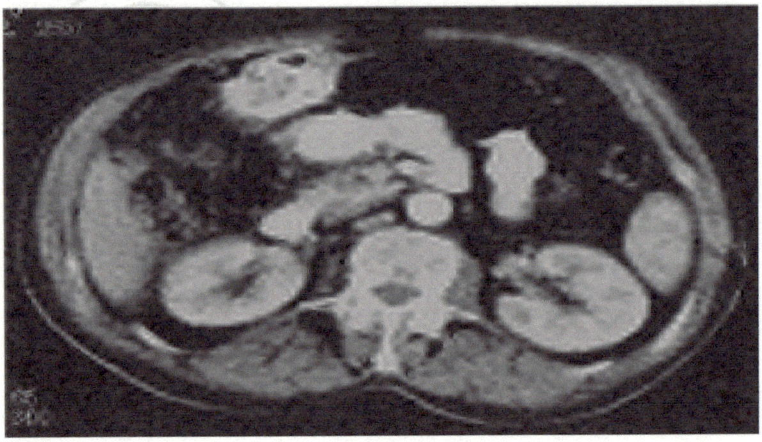

(b)

Figure 2.
An example of medical image denoising by applying thresholding in the DWT domain: (a) a noisy medical image with PSNR = 62 dB, (b) denoised image obtained from the application of a denoising technique based on thresholding in the DWT domain.

2.2 Image denoising

All digital images are degraded by different types of noise during their acquisition and transmission. As an example of these images, the medical one is likely disturbed by a complex sort of addition noise depending on the devices that are employed for capturing or storing it. There are no medical imaging devices that are noise free. The most commonly employed medical images are produced from MRI and CT equipment [1]. The additive noise corrupting medical image causes the reducing of the visual quality that complicates diagnosis and treatment.

Figure 2 illustrates an example of a medical image denoising using DWT.

A noise-added medical image and its denoised one obtained from employing a wavelet denoising technique are illustrated in **Figure 2**. The added noise has Gaussian distribution, and symlet 6, decomposition level of two, hard thresholding were used as the parameters for the application the wavelet-based denoising technique [1].

3. Conclusion

In this chapter, we deal with a number of signal and image denoising techniques existing in the literature. We also give two examples of signal and image denoising by applying the denoising techniques based on thresholding in the Discrete Wavelet Transform domain. Those examples show the performance of these denoising techniques.

Author details

Mourad Talbi
Laboratory of Nanomaterials and Systems for Renewable Energies (LaNSER), Center of Researches and Technologies of Energy (CRTEn), Tunis, Tunisia

*Address all correspondence to: mouradtalbi196@yahoo.fr

References

[1] Ergen B. Signal and image denoising using wavelet transform. In: Advances in Wavelet Theory and their Applications in Engineering, Physics and Technology. London, UK: InTech; 2012

[2] Chen G, Bui T. Multiwavelets denoising using neighboring coefficients. IEEE Signal Processing Letters. 2003; **10**(7):211-214

[3] Portilla J, Strela V, et al. Image denoising using scale mixtures of Gaussians in the wavelet domain. IEEE Transactions on Image Processing. 2003; **12**(11):1338-1351

[4] Buades A, Coll B, et al. A review of image denoising algorithms, with a new one. Multiscale Modeling and Simulation. 2006;**4**(2):490-530

[5] Moulin P, Liu J. Analysis of multiresolution image denoising schemes using generalized Gaussian and complexity priors. IEEE Transactions on Information Theory. 1999;**45**(3):909-919

[6] Alfaouri M, Daqrouq K. ECG signal denoising by wavelet transform thresholding. American Journal of Applied Sciences. 2008;**5**(3):276-281

[7] Wachowiak MP, Rash GS, et al. Wavelet-based noise removal for biomechanical signals: A comparative study. IEEE Transactions on Biomedical Engineering. 2000;**47**(3):360-368

[8] Kother Mohideen S, Perumal SA, Sathik MM. Image de-noising using discrete wavelet transform. IJCSNS International Journal of Computer Science and Network Security. 2008;**8**(1):2

[9] Liu D, Chen X. Image denoising based on improved bidimensional empirical mode decomposition thresholding technology. Multimedia Tools and Applications. 2019;**78**:7381-7417. DOI: 10.1007/s11042-018-6503-6

[10] Lahmiri S. Image denoising in bidimensional empirical mode decomposition domain: The role of Student's probability distribution function. Healthcare Technology Letters. 2016;**3**(1):67-71

[11] Akay M, Semmlow J, et al. Detection of coronary occlusions using autoregressive modeling of diastolic heart sounds. IEEE Transactions on Biomedical Engineering. 1990;**37**(4): 366-373

[12] Ergen B, Tatar Y, et al. Time-frequency analysis of phonocardiogram signals using wavelet transform: A comparative study. Computer Methods in Biomechanics and Biomedical Engineering. 2010;**99999**(1):1-1

Chapter 2

Sound Absorption Measurement: Alpha Cabin and Impedance Tube

Pavel Něměček

Abstract

The stage of development of absorbent materials, when necessary to verify their properties in relation to established requirements, plays one of the key challenges in current research. Nowadays, experimentation represents the only reliable way to quantify sound absorption. Thus, the determined sound absorption coefficient is used to compare individual development variants, and also, it is used in a selection of material from the commercial offer. Therefore, the main part of research is devoted to measurements in the impedance tube and in the alpha cabin, because these procedures play one of the most challenging roles in practice. All used experimental methods are based on the theory about transformation of sound energy into other forms of energy in the material. Nevertheless, the physical nature of sound absorption and individual measurement principles are not covered in this chapter, nor are any sound insulation measurements. It deals solely with the sound absorption and determination of the sound absorption coefficient. As a results, this chapter further summarizes basic information on a sound absorption measurement, and mainly, focuses on practical recommendations as well as applicability of results. First and foremost, these individual procedures may represent a considerable international overlap in the field.

Keywords: sound absorption coefficient, impedance tube, alpha cabin, sound absorption measurement, reverberation time

1. Introduction

Sound absorption measurement serves as an activity associated with the design, verification and application of suitable materials for solving an acoustic situation of enclosed spaces. These materials have a specific composition, may be applied to large surfaces, and their properties must typically meet many requirements (thermal insulation, mechanical resistance, low dirtiness, compactness, etc.). Absorbent materials are used in building acoustics, the automotive industry and everywhere where humans and noise sources are in a confined space.

This chapter covers the experimental determination of the sound absorption coefficient of industrially produced materials or samples in the stage of development, which are intended for the reduction, or regulation of noise in closed spaces. The chapter also contains a brief description of basic comparison methods and a more detailed description of two laboratory methods, i.e. the measurement of sound absorption in an impedance tube and in an alpha cabin. The goal is to provide a basic description of experimental methods, their comparison, evaluation and determination of accuracy.

In practice:

- the requirements for noise in a closed space are given,
- the properties of absorbent materials are derived from them,
- materials are developed,
- their properties are experimentally verified and compared with the requirements,
- materials are optimized,
- materials are applied,
- verification of the optimized space is carried out, compliance with the space requirements is evaluated.
- The results are put into practice.

2. Basics of measuring sound absorption

Sound absorption is the ability of a material environment to absorb coming sound. It is a process in which sound energy falling on a sample of material is transformed into another form, mainly thermal energy. In the ideal case, the sound energy that encounters the sample ($W_{INBOUND}$) is partly reflected ($W_{REFLECTION}$), partly transmitted through the sample ($W_{TRANSMISSION}$) and partly absorbed into the sample ($W_{ABSORPTION}$). Applies to:

$W_{INBOUND} = W_{REFLECTION} + W_{ABSORPTION} + W_{TRANSMISSION}$.

When experimentally investigating absorption, certain conditions need to be met so the equation above can be applied. The conditions are:

- It can be assumed that the material sample is fully involved in the energy balance.
- The source of sound energy is controlled.
- The measurement system does not significantly affect the measurement results (by its principle, dimensions and other metrological and non-metrological properties).

There are additional requirements for experimental methods:

- They should enable a statistical approach, respecting the variability of the measurement system as well as the variability within the sample.
- Selected method should satisfy the intended use of the results with its accuracy.

- Selected method should be fast and repeatable.
- The absorption frequency bandwidth should be as wide as possible.
- Method with a perpendicular incident of sound waves on the sample and
- Method with an omnidirectional impact of sound waves on the sample.

To the methods used, it should also be mentioned:

- Determination of sound absorption is always an estimation with a definable precision. A result cannot always be related to a specific application due to variability within samples (each sample taken from the production differs from another).
- A measurement is always indirect, the sound absorption coefficient is always calculated from other measured quantities.
- It is assumed for all measurement methods that the sample is placed near a surface that has zero sound transmission, hence $W_{TRANSMISSION} = 0$. Therefore, all incident acoustic energy is absorbed or reflected.
- Most of the methods compare results with and without a sample. Results without a sample assume absorption to be equal to zero.
- Variability caused by the instability of the excitation signal (energy encountering the sample) is averaged. Variability of the measurement system has a defined measurement uncertainty.

3. Physical and metrological basis of sound absorption measurement

A measurement can be considered as the only objective option to determine the sound absorption coefficient α. Other options, such as simulations and modeling in a virtual environment, face problems with an accurate determination of the boundary conditions and with the definition of the internal structure of the material. Validation of materials intended to solve a sound situation in closed spaces requires a determination of the sound absorption factor by experiment, therefore on a real part by objective methods. Requirements are set, for example, by the automotive industry or the building materials industry. Knowledge of the sound absorption coefficient makes it possible to model sound propagation in closed spaces using special software.

Measuring the sound absorption coefficient is a topic mainly for:

1. Independent testing laboratories,
2. Research organizations,
3. Industrial companies involved in the development and production of absorbent materials.

In the next part of the text, 4 measurement methods are described. Two of them are very simple and serve more for a comparison to the reference sample, the other two ones are the most used in practice. A method, which is used mainly by independent testing laboratories and complies the international standards be briefly mentioned.

The following points apply to all described methods:

1. Measurement methods are always indirect. The sound absorption coefficient is always determined by calculation from measurements of other quantities. The evaluation of sound absorption is based on the consequences of the energy conversion, which is a decrease in the sound pressure level (sound intensity) after passing through the sample, a change in the reverberation time after the application of an absorbing material or from the deformation of the reflected wave.

2. A known source(s) is used to excite the acoustic energy (incident acoustic energy).

3. An experiment assumes that the acoustic energy encountering the sample is partially absorbed by the sample and the rest of the energy is reflected. The sample therefore lies/rests on a soundproof surface during measurement. The measurement is always based on the principle that energy from the measuring system is prevented from passing through the sample. The assumption is that the incident energy is reflected from the surface behind the sample and partially absorbed as it passes through the sample. If a relatively large sample is available, the absorbed energy is evaluated using the reverberation time, for smaller samples the resultant wave (sum of incident and reflected wave) is mapped in the near space.

4. The result is a frequency spectrum of the sound absorption factor in the bandwidth allowed by the method.

5. The result is evaluated statistically.

3.1 Approximate and comparative methods of measuring sound absorption

In this chapter are described simple methods of sound absorption measurement. Their description serves rather to complement the technical and historical context of sound absorption measurements. It is based on conditions where it is not possible or preferred to use more advanced methods. These methods:

- Are easily applicable in operating conditions,

- Use a simple calculation model,

- Are relatively fast.

These methods cannot be characterized as laboratory methods for an objective determination of sound absorption. They only serve as quick comparison tests in the optimization of the composition of absorbing layers and can be used directly at the place of application.

3.1.1 Tone burst method

This method is currently used very rarely. However, in the available literary sources, it is still found in various methodological variants [1–3], it is popular with students who use it in a case when sound absorption is only one of the properties they investigate on materials.

This is a simple method based on the idea that when a sound wave strikes a sample at a certain angle, a reflected wave propagates at the same angle with energy reduced by the absorbed energy. The calculation is based on processing the ratio of incident energy and reflected energy.

The measurement takes place in a space that is as anechoic as possible, ideally using a directional microphone and a directional speaker.

1. Measurement step: a signal with sufficient energy and bandwidth is applied to the speaker and the sound pressure level $L_{p,d}$ (its frequency spectrum) is measured at the distance A between the speaker and the microphone.

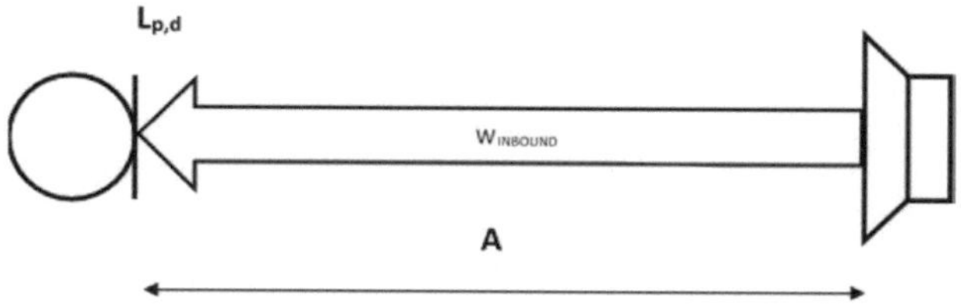

2. Measurement step: a testing object of sufficient dimensions is placed on the reflective pad. The distances from the microphone to the point of reflection of the sound waves and the distance of the loudspeaker from the point of reflection of the waves are identically A/2. The identical signal emitted by the speaker is partially absorbed in the absorbing wave, the rest is reflected to the microphone. The measurement takes place according to the diagram below, the important thing is the same angle of incidence/reflection φ. The frequency spectrum of the sound pressure level $L_{p,r}$ is evaluated at the microphone.

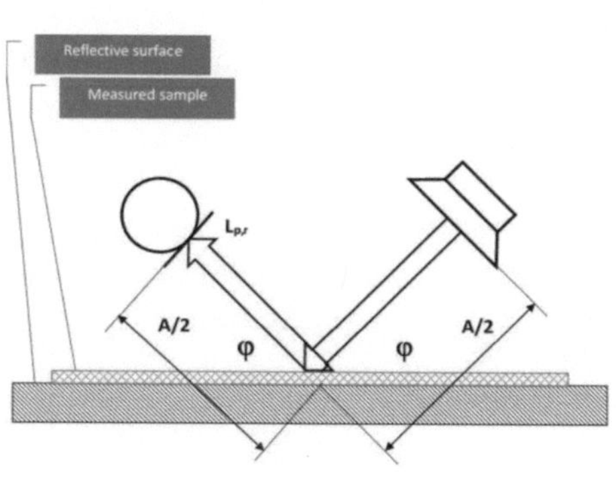

3. Step 3: at individual frequencies (1/3 octave), the sound absorption coefficient $\alpha(\varphi;f)$ is estimated according to the formula:

$$\alpha(\varphi;f) = 1 - 10^{-\frac{\left(L_{p,d}(\varphi;f) - L_{p,r}(\varphi;f)\right)}{10}} \quad (1)$$

Method Notes:

1. The estimated sound absorption coefficient is dependent on the angle φ. Although the method can be included in the group of methods with omnidirectional impact of sound waves, the dependence of the result on the angle of incidence/reflection φ is obvious.

2. For small angles φ, the method is very inaccurate.

3. The Tone Burst Method is very sensitive to background noise.

4. For example, glass can be recommended as a reflective surface, but the assumption that this surface does not transmit sound energy, which ultimately increases the resulting sound absorption, is not always valid.

3.1.2 Sound intensity measurement method

This approximate method requires the use of a measuring system with a sound intensity probe.

In an open field, the reflective surface is covered with a sufficiently large sample of the measured material. A sound source is placed at a sufficient distance from the surface of the sample. At a close distance from the sample (approx. 0.2 distance between the surface of the sample and the speaker), the average sound pressure level L_p and the average sound intensity level L_I reflected from the sample are measured. The sound absorption coefficient is then calculated by the formula:

$$\alpha(f) = \frac{4}{1 + 10^{\frac{L_P(f) - L_I(f)}{10}}} \quad (2)$$

The incident energy is proportional to the sound pressure level, the reflected energy is identified as a component of the sound intensity vector. Both quantities are identified by the sound intensity probe.

Advantages of the Sound Intensity Measurement Method may be:

- Low time consumption,

- The measurement does not take place in two steps without a sample/with a sample, but only with an applied sample, which gives the possibility to measure the absorbency of permanent applications.

The disadvantages of the Sound Intensity Measurement Method may be:

- Lower accuracy,

- The impossibility of relating the result only to the absorbent material sample. It is not possible to separate the absorption of the sample and the absorption of the substrate on which it is placed.

3.2 Accurate methods of measuring the sound absorption coefficient

3.2.1 Standard ISO 354

This international standard [4] defines the basic laboratory procedure for determining the sound absorption coefficient in a reverberation space. The procedure can be considered the most accurate procedure leading to the determination of the sound absorption coefficient. The method determines the sound absorption coefficient for diffusing the sound impact and can be used to measure materials with distinct shape structures in the straight and perpendicular direction. It is described in great detail in the standard and is especially suitable for specialized laboratories. The standard places strict requirements on the reverberation space, its dimensions and above all on the dimensions of the sample. The declared frequency range is from 100 Hz to 5000 Hz. The principle of indirect measurement of the sound absorption coefficient is based on Sabin's formula [5]:

A formula developed by Wallace Clement Sabine that allows designers to plan reverberation time in a room in advance of construction and occupancy. Defined and improved empirically the Sabine Formula [5] is.

$$T(60) = 0.161 \cdot \frac{V}{A} |s|, \tag{3}$$

Where:

T(60) = reverberation time or time required (for sound to decay 60 dB after source has stopped) |s|,

V = Volume of room $|m^3|$,

A = the equivalent absorption surface $|m^2|$.

In the test room, the reverberation time is measured with and without the mounted test sample. The reverberation time is the time during which the sound pressure level decreases by 60 dB after the sound source is turned off. This means that the original acoustic energy drops to 1/1000000 of its original size. In the test room, the reason for the decrease is the sound absorption and then the reverberation time is its measure. The equivalent surface is a hypothetical surface of a perfectly absorbing sample that has the same properties as a real sample. The equivalent area is the basis for calculating the sound absorption coefficient.

Advantages of determining the sound absorption coefficient according to ISO 354:

- the possibility of measuring samples with large thicknesses,
- accuracy, repeatability,
- international acknowledgement.

Disadvantages of determining the sound absorption coefficient according to ISO 354:

- Reverberation room volume of at least 150 m^3,
- Requirements for the shape of the room,
- Sample area 10 m^2 to 12 m^2,
- Possibly higher price for the testing.

Above all, the requirement for a size of a sample is a problem of using this basic method in the sample development phase, when many possible variants are experimentally verified with subsequent optimization. It is practically impossible for manufacturers of absorbent materials and research organizations that are not directly oriented towards this research to acquire such expensive laboratory facilities.

Specific information on the measurement and calculation procedure is contained in the mentioned standard and it is not the aim of this chapter to discuss them in more detail.

3.2.2 Measurement in an impedance tube

An impedance tube is the most common device used today to estimate sound absorption. In the professional literature, this method is currently mentioned most often. There are more concrete technical versions of the tube, from the own construction of a research workplace to commercially offered versions. As an example, **Figure 1** shows the assembly from Brüel & Kjær Impedance Tube Kit 4206 (4206-A), which is described in the following text.

The impedance tube principle is based on the creation of a combination of direct and reflected waves in a rigid closed tube with an internal smooth and reflective surface. The skeleton of the tube must be as soundproof as possible. One end of the tube covers a sample that is being measured, on the other end of the tube there is placed a speaker that excites by broadband noise the inner volume of the tube. A plane wave is created between the speaker and the sample, which is a combination of incident and reflected waves. The energy of the reflected wave is reduced by the energy absorbed in the sample. The resulting wave is sampled in the tube and an estimate of the sound absorption coefficient is determined by evaluating the data obtained. The sound wave strikes the sample perpendicularly.

Basic characteristics of impedance tube measurement:

- A dimensionally small sample is measured (100 mm/39 mm for [6]), which is an advantage for the development and optimization of materials, but a disadvantage if the developed materials have significant spatial elements, change on the surface or contain significant non-homogeneities.
- The strike of the sound wave is perpendicular to the sample and the measurement result corresponds to this. The results therefore do not correspond to the behavior of the material in real conditions, where the omnidirectional impact of sound waves prevails. The impedance tube is therefore particularly suitable as a precise comparison platform for the development of absorbing materials.
- The detected sound absorption coefficient represents the minimum ability to absorb sound and it can be assumed that the results will be better with

Figure 1.
Impedance tube and absorptive material samples (author's archive).

the methods in the diffusion field according to ISO 354 [4] and in the alpha cabin [7, 8].

- The methods are very sensitive to sealing a sample in a tube so that all acoustic energy passes through or reflects off the sample.
- If measurements are made for different sample diameters, it is necessary to unify the results in common frequency bands (principally by averaging).
- If a comprehensive idea of the sound absorption is to be obtained, it is advised to take and measure several samples from the research batch in order to cover all possible non-homogeneities and shape and material changes.
- Despite all the disadvantages, the impedance tube is a suitable and most commonly used platform for estimating the sound absorption coefficient and an important aid for the development of absorbing materials, mainly due to the speed of measurement and sample size.

3.2.2.1 Method using standing Wawa ratio

This method determines the sound absorption coefficient of acoustic materials when the sound is incident perpendicularly. The specific procedure for determining the sound absorption coefficient is described in [9]. The absorbing sample is fixed at one end of the tube. An incident plane sine wave p_i is excited by a speaker at the opposite end of the tube. By superposition $p = p_i + p_r$ of the pressures of the incident wave p_i and the reflected wave p_r from the test sample, a standing wave is created in the tube. The course of the sound pressure level of this standing wave is measured by an adjustable microphone, which is moved along the axis of the tube through the hole in the center of the speaker. The evaluation of sound absorption is based on the difference in sound pressure levels ΔL between the pressure maximum and minimum in the tube.

$$\alpha = \frac{4 \cdot 10^{\Delta L/20}}{\left(10^{\Delta L/20} + 1\right)^2} \tag{4}$$

Moving the microphone and accurately identifying the maximum and minimum sound pressure level reduces the speed of the sound absorption coefficient measurement. Impedance tubes for this evaluation method are more often an individual product of test laboratories, which allows adaptation to the desired frequency band and the way of moving the microphone and evaluating the absorption.

3.2.2.2 Transfer-function method

This test method is similar to the previous method in that it uses the same experimental scheme with a sound source at one end and a sample fixed in an impedance tube at the other end. The procedure is described in detail in [10, 11]. In this test method, plane waves in the tube are excited by a noise source and the sound pressure is measured by microphones located at two fixed points (or by one microphone moved in the tube) and by subsequent calculation of the complex transfer function at a perpendicular incidence of sound waves. The test method is overall much faster than the measurement procedure described in the previous chapter.

The test sample is fixed to one end of a straight, rigid, smooth and sealed impedance tube. Plane waves are excited in the tube by a sound source (noise) and the sound pressure is measured by microphones at two locations near the sample. A complex transfer function is determined from the measured signals, which is used to calculate the sound absorption coefficient. The frequency range of the measurement depends on the dimensions of the tube and the distance between the positions of the microphones. In order to determine the sound absorption coefficient in a wider frequency range, measurements are made on an assembly that contains tubes of two different diameters. **Figure 1** shows a measuring set-up that allows determining the sound absorption coefficient for a sample diameter of 100 mm in the frequency range of 50 Hz to 6.4 kHz (for a sample thickness of 440 mm maximum [6]) and for a sample diameter of 29 mm in the frequency range of 100 Hz to 3.2 kHz (for a sample thickness of 200 mm maximum [6]).

Measurements can be done:

1. By a method using two microphones that simultaneously measure the signal in the tube at two clearly defined points,

2. By a method of one microphone, which is moved gradually to two measuring locations during the measurement.

Procedure 1 is quick, accurate and easier to do. It is widespread in practice and much more published.

Procedure 2 requires a specialized excitation signal, has more demanding requirements for processing the measured signals, and is more time-consuming. It better eliminates phase mismatch between microphones and allows optimal selection of microphone locations for each measured frequency. According to [10], this procedure is recommended for evaluating of tuned resonators.

Advantages of measuring in an impedance tube:

- small sample size,
- fast measurement,
- relatively available measuring technology, availability of laboratories,
- strong information and publication background,
- the existence of an international standard.

Disadvantages of measuring in an impedance tube:

- assessment of absorption only for the perpendicular impact of sound waves,
- limited sample thickness.

3.2.3 Measurements in the alpha cabin

The Alpha cabin [7, 8, 12, 13] is an internationally acknowledged measurement platform for determining the sound absorption coefficient at the omnidirectional impact of sound waves. It is therefore close to measurements according to ISO 354, it is based on the requirements of this standard, it respects the methodology as much as possible, but removes the disadvantage of the need for large samples.

The Alpha cabin is a platform that is scaled 1:3.2 to the echo chamber parameters of the Swiss Material Testing and Testing Laboratory (EMPA) in Dübendorf. **Figure 2** shows an example of the latest design of the alpha cabin. It is a reverberant space sound-isolated from the outside environment with non-parallel walls.

The main technical data of the alpha cabin are:

Internal cabin volume:	6.44 m^3
Frequency measurement range:	400 Hz to 10 kHz (1/3oct.)
Dimensions of a standard sample:	1.0 m × 1.2 m
Surface of absorbent parts:	0.6 m^2 to 2.4 m^2

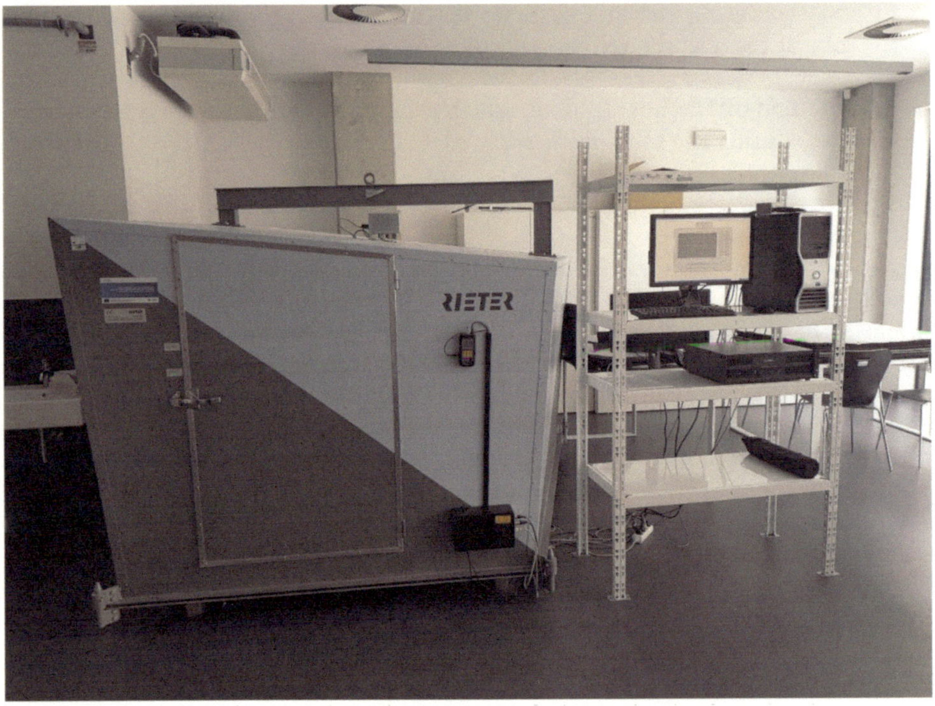

Figure 2.
Alpha cabin - Technical University of Liberec (author's archive).

The formula [7] is used to determine the sound absorption coefficient:

$$\alpha_S = \frac{0{,}966}{S}\left(\frac{1}{T_1} - \frac{1}{T_0}\right) \tag{5}$$

Where the measured quantities are:
S = sample area $|m^2|$.
T_1 = reverberation time in the sample booth |s|,
T_0 = reverberation time in the cabin without sample |s|.

The ratios in the diffusion field of the alpha cabin (**Figure 3**) are practically the same as in the large reverberation chamber, but for three times shorter wavelengths (three times higher frequencies). The Alpha cabin therefore provides comparable results on much smaller sample areas than required by ISO 354. However, the proportional changes in cabin conditions run into one problem. The thickness of the sample is the only geometric quantity that cannot be reduced in a ratio of 1:3, and thus the absorbing surface corresponding to the edges of the sample appears three times larger in proportion to its surface. The problem must be eliminated by edging the side surface of the sample with soundproof material. As standard, it is solved by a metal bounding frame with the dimensions of a standard sample, which is higher than the usual thicknesses of the developed materials. In the case of larger thicknesses, it is recommended to manufacture your own frame and validate it using a reference sample. The importance of sample edging can be shown on the measurement results of the same sample with a thickness of 20 mm with and without a bounding frame in **Figure 4**. The course shows that the error of determining the sound absorption coefficient increases with increasing frequency (**Figure 5**).

DOI: http://dx.doi.org/10.5772/intechopen.110410

Figure 3.
Alpha cabin with embedded and framed sample (author's archive).

The Alpha cabin measurement procedure generally requires two measurements.

1. Determination of the reverberation time in a cabin with a frame without a sample T_0,

2. Determination of the reverberation time in the cabin with a framed sample T_1.

The sound absorption coefficient is then calculated according to formula (5).

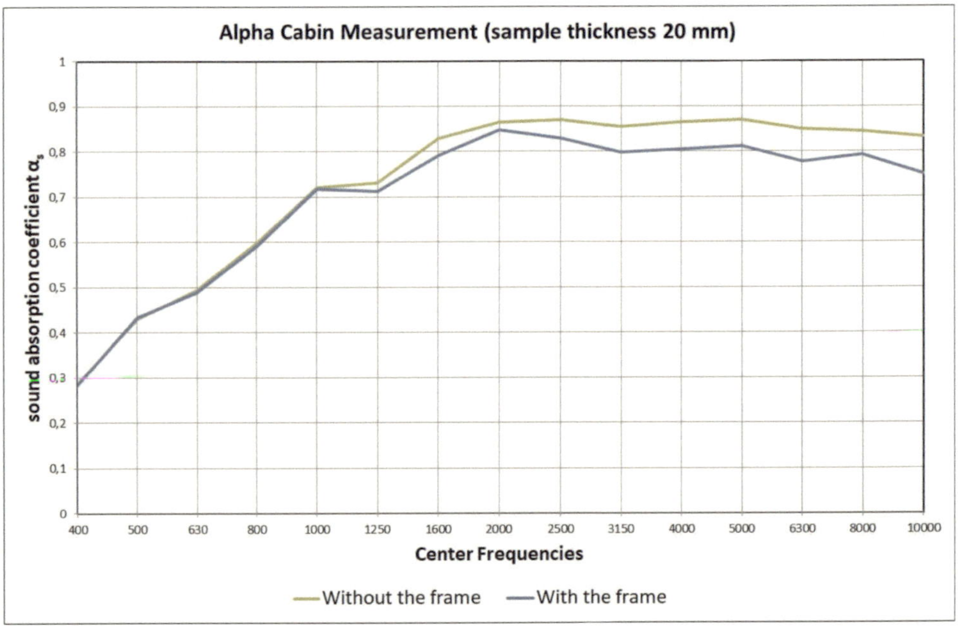

Figure 4.
Results of measurements with and without a border frame.

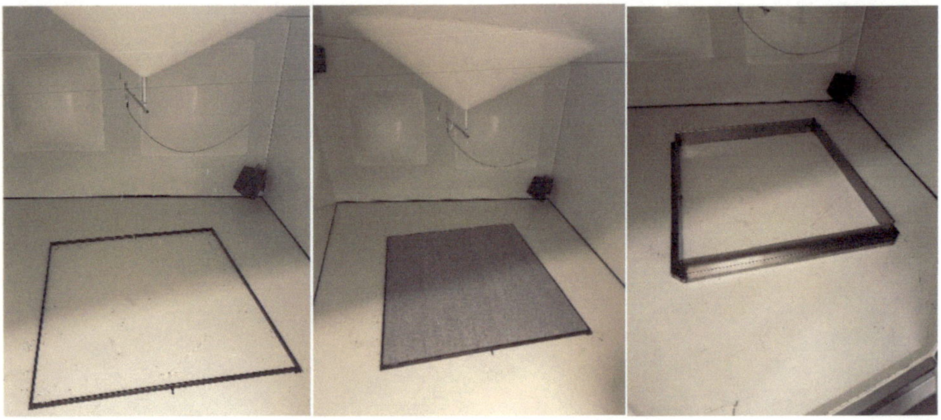

Figure 5.
Frames for delimiting samples (author's archive).

The Alpha cabin has one more function, which is the evaluation of absorbing objects. These are shaped parts that absorb sound, but the absorbing surface cannot be determined. A typical example is the seat of a passenger car [12], which significantly affects the noise in the closed space of the cabin due to its absorption, but it is not possible to clearly determine the absorbing surface, or to create a sample of standardized dimensions from the seat. In Eq. (5) it is not possible to substitute the absorbing surface S, and thus the measurement result is equal to the equivalent absorbing surface A:

$$A = 0{,}966 \cdot \left(\frac{1}{T_1} - \frac{1}{T_0}\right) \qquad |m^2| \tag{6}$$

The equivalent absorptive surface corresponds to the absolute absorptive surface ($\alpha_S = 1$), which has the same absorptive capacity as the shaped part. Therefore, the larger the equivalent surface area, the more the shaped part is able to absorb more sound energy. An example of the result of measuring the equivalent absorbing surface for a shaped part in the construction of a passenger car is shown in **Figure 6**.

The equivalent absorptive surface of shaped parts is primarily a comparative parameter when developing or selecting a part for a protected space. However, it can be used in the calculations of the total absorption, because according to Eq. (5):

$$A = S \cdot \alpha_S \tag{7}$$

Advantages of measuring sound absorption in the alpha cabin:

- Measurement of the sound absorption coefficient for the omnidirectional impact of sound waves,
- Optimal sample size considering the possibilities of developing absorbing materials and the accuracy of the estimate,
- The possibility of determining the equivalent absorption A of shaped parts,
- Test speed and repeatability,
- A recognized platform in the automotive industry,

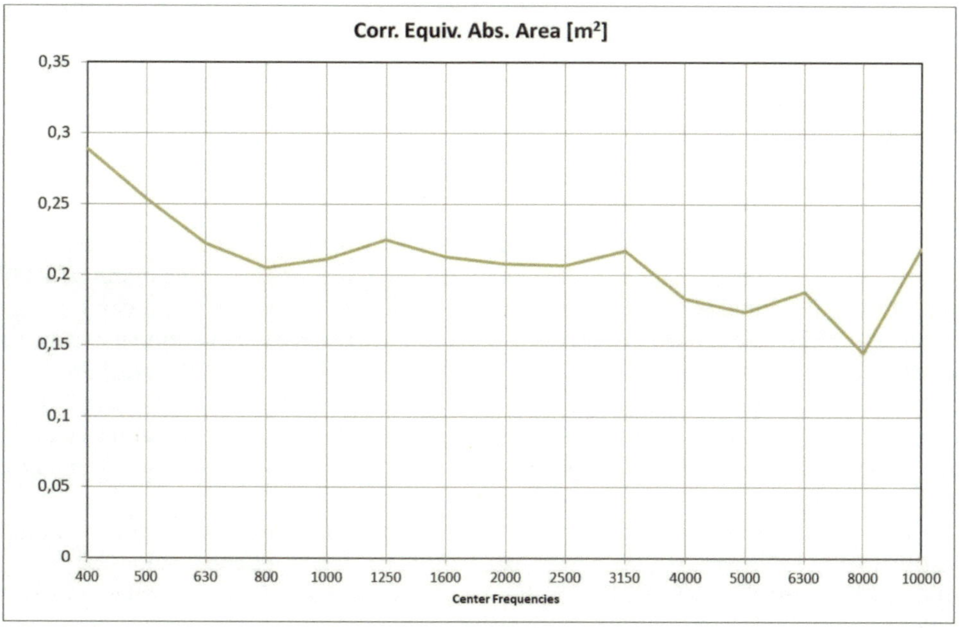

Figure 6.
Measurement results of the equivalent absorptive surface of a car component.

- Possibility of estimating absorption also for non-homogeneous materials (recycled materials, disordered fibrous materials, loose substrate, etc.),
- Possibility of estimating the absorption of decorative panels and artworks designed to solve the acoustics of closed spaces.

Disadvantages of measuring sound absorption in the alpha cabin:

- relatively large size of the sample compared to the impedance tube,
- limited sample thickness due to the need for lateral sealing,
- large and expensive measuring equipment.

4. Conclusion

Measurement is still the most accurate and fastest procedure for determining the sound absorption coefficient. The practical use of absorbing materials in practice requires an objective determination of absorption for the purpose of optimizing the acoustic properties of enclosed spaces.

The current development of absorbent materials is predominantly still using fibers or porous raw materials with an emphasis on other important properties, such as ecology, usability of waste and recyclable resources, esthetics, non-flammability, etc. For products designed in this way (mats, panels, absorbent elements) the main principle of absorption is the conversion of sound energy into heat by friction of the internal structure of the absorbing element. In general, the elements then have optimal efficiency starting from the frequency that is determined by following equation:

$$f = \frac{86\,000}{H} \ |\mathrm{Hz}| \tag{8}$$

Where:
H = the thickness of the absorbing element |mm|,
f = frequency |Hz|.

The optimal thickness equals to a quarter of the wavelength of a perpendicular incident wave, so it can be considered the minimum value at which the material is able to use its full potential to absorb sound. With omnidirectional impact, it can be assumed that the optimal bandwidth will shift to lower frequencies. **Figure 7** shows an example of the measurement result in the Alpha cabin of a sample of absorbent material with a reference thickness of 22 mm. A frequency of 3.9 kHz (4 kHz 1/3 octave) corresponds to a thickness of 22 mm. Due to the omnidirectional impact of sound waves, the maximum absorption value is maintained even at lower frequencies (2 kHz).

There is an inverse proportional relationship between the optimal frequency and the thickness of the material. When considering declared frequency ranges of individual measurement methods, **Figure 8** provides a comprehensive overview of the methods and their practical use for materials testing.

It is clear from **Figure 8** that the lowest declared measurement frequency achievable in the impedance tube is at frequency of 50 Hz. This would correspond to the

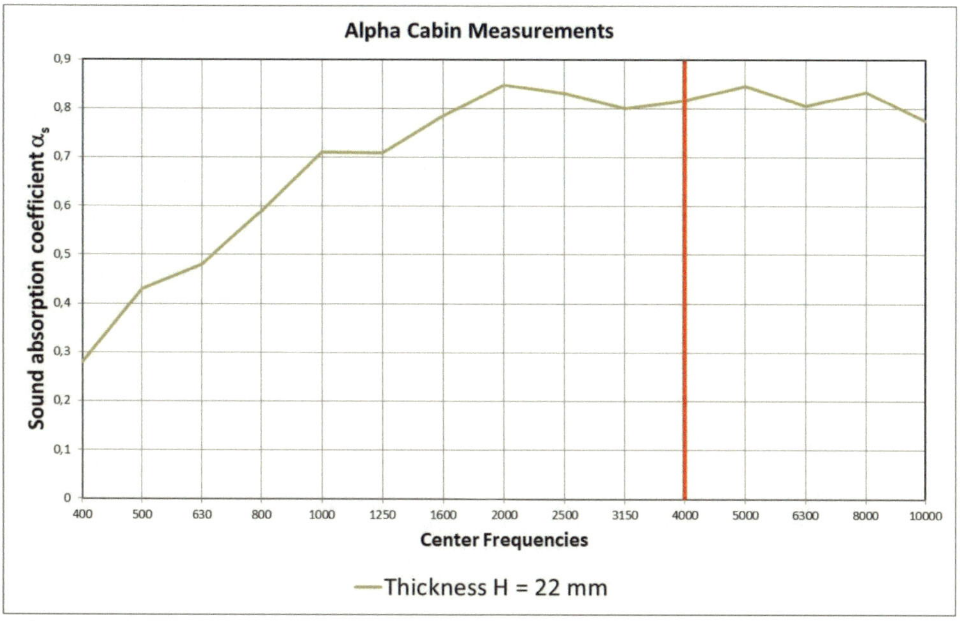

Figure 7.
Sound absorption of a sample with a thickness of 22 mm - measured in the alpha cabin.

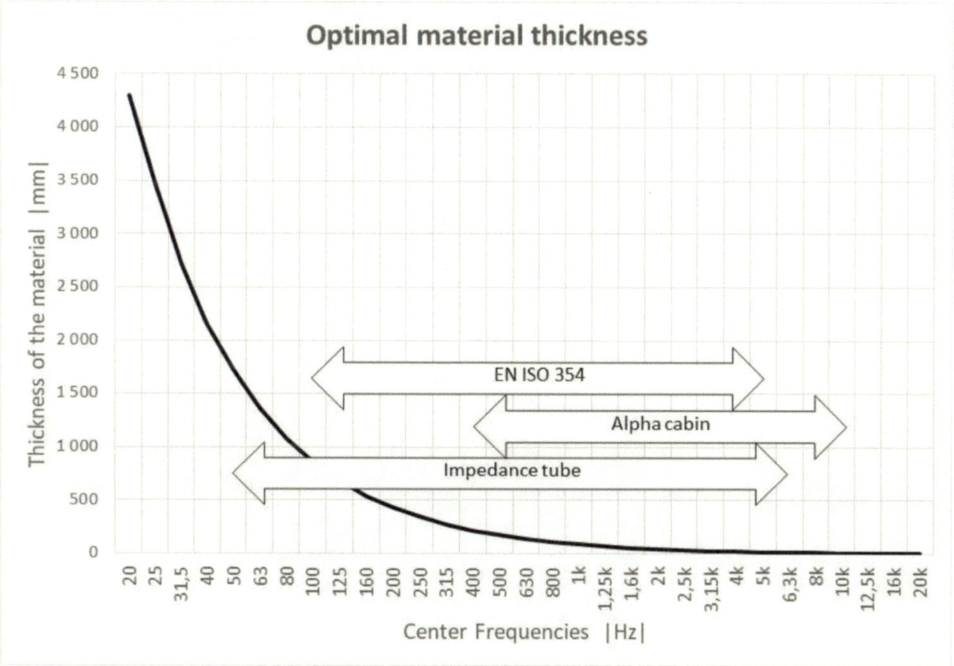

Figure 8.
Measurement methods in relation to the optimization of the thickness of the absorbent material.

optimal absorption of materials with a thickness of approx. 1700 mm, which is technically impossible. An impedance tube of such dimensions is not used in practice, but a sufficiently wide frequency band is available for an objective assessment of commonly used materials. The Alpha cabin starts at a frequency of 400 Hz, which corresponds to roughly 200 mm of material thickness when optimally used. This thickness is usable for the Alpha cabin. The ISO 354 standard declares a minimum frequency of 100 Hz, which corresponds to 860 mm of optimal thickness, which is also acceptable given the dimensions of the space and the area of the sample. It should be emphasized that the measurement according to the ISO 354 standard and in the Alpha cabin is based on the omnidirectional impact of sound waves, the impedance tube is based on only a perpendicular impact.

From the **Figure 8**, an uncovered bandwidth of sound absorption measurements up to 100 Hz can be seen in the case of omnidirectional impact of sound waves. It should be emphasized that physical and technical obstacles to the use of independent methods are encountered here. The optimal thickness of the materials is greater than 860 mm and ends at 4.3 m for 20 Hz, which is the lowest frequency of the audible band. However, this range of thicknesses of absorbing materials is difficult to use in the real world for practical reasons. The exception is specialized anechoic chambers with high volumes. Here, the effectiveness of absorbing materials is assessed by measuring the reverberation time directly during implementation.

If the commercially usual area of absorbent materials (up to a maximum thickness of 200 mm) were to be evaluated, it can be seen from **Figure 9** that the optimal platform is the alpha cabin.

Recommendations for the design and experimental verification of the properties of absorbent materials

If the absorbing material is to fulfill the expectations, its structure must be properly designed. This is a matter of material development respecting other requirements (legislative requirements, esthetic requirements, applicability in specific conditions,

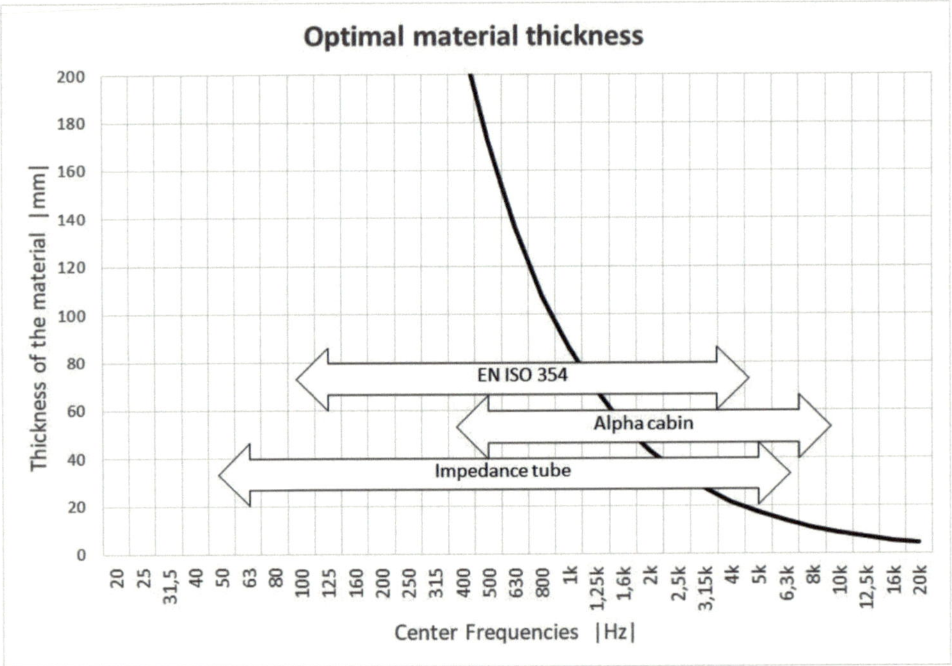

Figure 9.
Measurement methods in relation to thickness optimization of conventional absorbent material / author's archive /.

other specific customer requirements). An objective assessment of sound absorption can only be achieved by measuring on an existing sample. Below is the basic procedure for determining the sound absorption coefficient by experiment.

1. To perform the measurement, a sample must be taken that clearly states:

    - what stage of development it is in, whether it is a measurement for validation or for comparison,

    - how should it be applied (directly on the wall, with an air gap, in combination with other materials)?

2. It is necessary to establish what are the requirements for the accuracy of the sound absorption coefficient measurement result. The accuracy of the result is influenced by the combination of measurement speed, used HW and SW, adherence to sample size, variability of measurement conditions and adherence to methodology. If it will be a comparative measurement that is carried out using the same method, a lower accuracy of the result can be accepted. If it is a question of describing the final form of the absorbing material, it is advisable to recommend methods with a higher quality. Standard ISO 354 can be considered the method with the highest accuracy.

    Lower accuracy methods:

    - Tone Burst Method

    - Sound Intensity Measurement Method

    Higher accuracy methods:

    - ISO 354 standard

    - Measurement in an impedance tube

    - Measurements in the alpha cabin

3. The sound absorption coefficient is determined by the selected method. It is advisable to apply a statistical approach, perform repeated measurements, determine the result and its uncertainty.

4. It is evaluated whether the measurement result met the expected goals.

5. Discussion

This chapter summarizes experimental methods for determining the sound absorption coefficient α. The chapter addresses the user (researcher, customer, project solvers) who are tasked with designing (developing) absorbing material and need to verify it during the development stage or after application. Available approximate and exact methods, their starting points, limitations, advantages and disadvantages and usability in practice are described. The solver can thus choose a suitable method for the individual stages of the project solution or correctly formulate requirements for external laboratories. The chapter further helps to understand the measurement results in relation to the application to a specific space and guides the project solver to be aware of possible limitations and problems.

Recommendation:

1. The most accurate measurement of the sound absorption coefficient is according to the ISO 354 standard, but it requires a large sample area and expensive measuring equipment. It is not suitable for the phase of development and optimization of the properties of absorbent materials.

2. Objective results for practical application are provided by the alpha cabin, which optimally combines sample size and measurement accuracy. However, the cost of measuring equipment is relatively high.

3. For the phase of development and optimization of properties, an impedance tube is ideal, which requires a minimum sample size, but the results are only suitable for comparing individual variants.

4. Approximate methods are suitable when sound absorption is not an essential property or it is necessary to estimate the properties of an already applied material.

5. For material design and evaluation of sound absorption, it is necessary to have at least basic knowledge in the field of acoustics.

6. Sound absorption is dependent on sound frequency and optimal absorption can be achieved for material thicknesses according to Eq. (8).

7. Sound absorption and sound insulation are different properties of a material.

Author details

Pavel Němeček
Technical University of Liberec, Liberec, Czech Republic

*Address all correspondence to: pavel.nemecek@tul.cz

References

[1] Brüel & Kjær. Measurements in Building Acoustic [Internet]. 1988. Available from: https://www.academia.edu/17353265/Measurementsin_Building_Acoustics?email_work_card=view-paper

[2] Bérengier M, Garai M. A State-of-the-Art of in situ Measurement of the Sound Absorption Coefficient of Road Pavements. Available from: https://www.academia.edu/34047160/MEASUREMENT_METHODS_OF_ACOUSTICAL_PROPERTIES_OF_MATERIALS_SESSIONS_A_state_of_the_art_of_in_situ_measurement_of_the_sound_absorption_coefficient_of_road_pavements

[3] Quintero-Rincón A. Measurement of the Sound-Absorption Coefficient in situ in Eggs Cartons using the Tone Burst Method. Available from: https://www.academia.edu/12996142/Measurement_of_the_sound_absorption_coefficient_in_situ_in_eggs_cartons_using_the_Tone_Burst_Method?email_work_card=title

[4] ISO 354, Acoustic - Measurement of Sound Absorption in a Reverberation Room. Switzerland; International Organization for Standardization; 2003

[5] Reverberation Time and Sabine's Formula. Available from: https://www.acousticlab.com/en/reverberation-time-and-sabines-formula/

[6] Product Data. Impedance Tube Kit (50 Hz – 6.4 kHz) Type 4206, Impedance Tube Kit (100 Hz – 3.2 kHz) Type 4206-A, Transmission Loss Tube Kit (50 Hz – 6.4 kHz) Type 4206. Available from: https://www.bksv.com/media/doc/bp1039.pdf

[7] Alpha Cabin. A Global Standard for Acoustic Absorption Measurements. Available from: https://www.autoneum.com/alpha-cabin-ii/

[8] Alpha Cabin. Available from: http://www.cttm-lemans.com/en/acoustics-vibrations/produits/alpha-cabin.html

[9] ISO 10534-1, Acoustics - Determination of Sound Absorption Coefficient and Impedance in Impedance Tubes - Part 1: Method using Standing Wave Ratio. Switzerland: International Organization for Standardization; 1996

[10] ISO 10534-2, Acoustics – Determination of Sound Absorption Coefficient and Impedance in Impedance Tubes – Part 2: Transfer-Function Method. Switzerland: International Organization for Standardization; 1998

[11] Yaochi Tang Y, Chuang X-J. Tuning of Estimated Sound Absorption Coefficient of Materials of Reverberation Room Method. 2022. Available from: https://www.hindawi.com/journals/sv/2022/5192984/

[12] Megasorber, Alpha Cabin. Acoustic Performance, Sound Absorption Test Chamber. Available from: https://megasorber.com/alpha-cabin/

[13] Nemecek P. Sound Absorption Measurement in Alpha Cabin. 2014. Available from: https://link.springer.com/chapter/10.1007/978-3-319-05203-8_52

Section 2

Image Denoising

Chapter 3

SAR Image Denoising Using MMSE Techniques

Mohamed Yahia and Tarig Ali

Abstract

Synthetic aperture radar (SAR) provides many advantages over optical remote sensing, principally the all-weather and all-day acquisition capability. For this reason, SAR images have been exploited for many applications such as forestry, agriculture, disaster monitoring, sea/ice monitoring. However, the main limitation in SAR images is the contamination with the multiplicative speckle noise. The speckle damages the radiometric quality of SAR images and contracts the performance of information extraction techniques. Many methods have been proposed in the literature to reduce speckle noise. These methods, however, must avoid degrading the useful information in the SAR images, such as textures, local mean of backscatter, and point targets. The minimum mean square error (MMSE) techniques have been largely applied in SAR image speckle denoising. The objective of this chapter is to review and give new insights into the MMSE denoising of SAR images. In particular, the performances of three MMSE-based techniques which are the commonly applied Lee sigma filter and the newly introduced iterative MMSE (IMMSE) filter, and the infinite number of looks prediction (INLP) filter are studied.

Keywords: SAR image denoising, MMSE techniques, Lee sigma filter, IMMSE filter, INLP filter

1. Introduction

Remote sensing imagery constitutes nowadays an important source of information for the characterization of the Earth's surface. The potentiality of synthetic aperture radar (SAR) systems is recognized for geoscience and remote sensing applications due to their operation in all-time and all-weather conditions. However, due to the coherent nature of the scattering mechanisms, SAR data are affected by the multiplicative speckle noise. The presence of speckle noise disturbs human interpretation of the images and reduces the accuracy of postprocessing such as image classification [1].

The multi-looking process (i.e., boxcar filter) reduces speckles by averaging the intensities of neighboring pixels [2]. Nevertheless, the spatial resolution is degraded. Many other denoising techniques have been introduced in the literature to alleviate this limitation by using other estimation domains including spatial [3] and wavelet [4]. In the intensity-driven adaptive-neighborhood (IDAN) filter a region-growing technique is applied to produce an adaptive neighborhood [5]. The total variation (TV) techniques [6] have been widely applied for SAR image denoising due to their efficiency to preserve spatial details and speckle reduction. The nonlocal NL filtering represents one of the powerful speckle reduction techniques. Zhong et al. [7] applied the NL Means (NLM) to filter SAR images by adapting the use of Euclidean distance to multiplicative noise. The probabilistic patch based (PPB) filter introduces a patch-based weight to generalize the Euclidean distance-based weight used in the NL means algorithm [8]. A hybrid NL-wavelet domain denoising technique has been proposed [9]. Penna et al. replaced the Euclidean distance in the NLM filter with stochastic distances in the Haar wavelet domain [10]. The NL-based filters improved significantly the denoising performance of SAR images. However, their main disadvantage resides in the high computing cost. Deep Learning techniques constitute a recent trend of PolSAR speckle filtering [11–13].

The minimum mean square error (MMSE) based filters that account for the local statistics of the image constitute an important branch of speckle filtering techniques. Since the introduction of the Lee sigma filter in early 1980 [14, 15], many versions have been elaborated such as Frost [16], Kuan [17], the improved Lee [18, 19], etc. Due to their effectiveness in speckle reduction, simplicity and low computational demand, many MMSE-based filters have been implemented in remote sensing software. To mitigate the drawbacks of the Lee sigma filter, various versions of the iterative MMSE (IMMSE) filter have been introduced recently [20–26]. Based on the MMSE principle, it has been demonstrated that the filtered pixels and their variances are linearly related. Then, a linear regression of means and variances for different window sizes is applied to estimate the infinite number of looks prediction (INLP) filtered pixels [27–30]. In this chapter, the improved MMSE-based Lee sigma, the IMMSE, and the INLP denoising techniques are studied.

This paper is organized as follows: Section 2 reviews the classical MMSE-based denoising technique and presents the updated versions, i.e., INLP and IMMSE techniques. The results are shown in Section 3. Finally, Section 4 presents the conclusions of this paper.

2. MMSE-based filters: classical version

The intensity pixel $y(i)$ of a SAR image is affected by a multiplicative noise [2]

$$y(i) = x(i)\,\nu(i), \tag{1}$$

$x(i)$ is the noise-free pixel and $\nu(i)$ is the speckle noise with unit mean and standard deviation σ_ν. It is assumed that $x(i)$ and $\nu(i)$ are statistically independent. In the rest of the chapter, the index (i) will be omitted. Let $\hat{x}$ and $\overline{x}$ be the estimated and the *a priori* mean of x, respectively.

From (1) we have

$$E(y) = E(x\,\nu) = E(x)\,\mathrm{E}(\nu) = E(x) \tag{2}$$

where $E()$ is a mathematical expectation (i.e., statistical mean). By exploiting the ergodicity of the SAR data, the statistical mean is substituted by the spatial mean, i.e.,

$$\overline{y} = \mathrm{E}(\mathrm{y}) = \mathrm{E}(\mathrm{x}) = \overline{x} \tag{3}$$

Nevertheless, it has been demonstrated recently that the statistical and spatial averaging statistics are quite different [27] since in the spatial averaging process the processing windows are overlapping, and a spatial correlation is introduced. Hence, replacing the statistical mean with the spatial one should be taken with caution.

The MMSE filter is assumed to be a linear combination of x and $\overline{x}$ [1]

$$\hat{x} = a\overline{x} + by \tag{4}$$

The parameters a and b are selected optimally to minimize the MSE

$$I = E\left[(x\text{-}\hat{x})^2\right] \tag{5}$$

then

$$\frac{\partial I}{\partial a} = 0 \tag{6}$$

and

$$\frac{\partial I}{\partial b} = 0 \tag{7}$$

This minimization leads to [1]

$$a = 1 - b \tag{8}$$

and

$$b = \frac{\mathrm{var}(x)}{\mathrm{var}(y)} \tag{9}$$

For a given X and Y two independent random variables, we have

$$\mathrm{var.(XY)} = \mathrm{E(Y)^2 var(X)} + \mathrm{var.(Y)var.(X)} + \mathrm{var.(Y)E(X)^2} \tag{10}$$

Then

$$\mathrm{var}(\mathrm{y}) = \mathrm{var}(\mathrm{x}\,\nu) = \overline{x}^2\sigma_v^2 + \mathrm{var}(\mathrm{x})\,\sigma_v^2 + \overline{v}^2\mathrm{var}(x) \tag{11}$$

Eq. (2) gives

$$\overline{y} = \overline{x} \tag{12}$$

then

$$\mathrm{var}(x) = \frac{\mathrm{var}(y) - \overline{y}^2\sigma_\nu^2}{1 + \sigma_\nu^2} \tag{13}$$

finally

$$\hat{x} = \overline{y} + b(y - \overline{y}) \tag{14}$$

$$b = \frac{\text{var}(y) - \bar{y}^2\sigma_v^2}{\text{var}(y)\left(1 + \sigma_v^2\right)} \tag{15}$$

var(y) and $\bar{y}$ are estimated using a moving window W. σ_v^2 is assumed to be a constant (i.e., $\sigma_v^2 = 1$ for single look SAR data).

2.1 Lee sigma filter

The Lee sigma filter has been implemented in several geographic information system (GIS) software due to its effectiveness in speckle reduction, its simplicity, and its computational efficiency. However, in amplitude and intensity SAR data, the probability density functions (pdf) are not symmetrical, because they follow the Rayleigh and the negative exponential distributions, respectively. This asymmetry produces biased estimates since the original sigma range was derived based on Gaussian distribution. Hence, to remove the bias and to preserve the mean value, the sigma ranges were recomputed based on the corresponding pdf. The sigma ranges of amplitude and intensity SAR data are given in [18]. In [28], the performance of the improved Lee sigma filter is revised.

Practical implementation

i. Define a square window W

ii. Define σ_v^2 ($\sigma_v^2 = 1/\text{N}$ where is the Initial number of looks (N = 1 in our study))

iii. Compute the statistics of the pixel ($\bar{y}$ and var(y))

iv. Compute the parameter b using (15)

v. Compute the filtered pixel $\hat{x}$ using (14).

3. MMSE-based filters: New insights

3.1 The IMMSE filter

The objective of SAR speckle filtering is:

- Reduce speckle noise in the homogeneous areas (i.e., averaging all pixels $\hat{x} = \bar{y}$).
- Maintain the spatial details (i.e., $\hat{x} = y$).

Hence in general cases, we have.

$$\hat{x} \in [y, \bar{y}] \text{ or } \hat{x} \in [\bar{y}, y] \tag{16}$$

The principle of the IMMSE filter is to scan the range of $\hat{x}$ in $[y, \bar{y}]$ by the following iterative procedure.

$$\hat{x}_0 = \bar{y}, \tag{17}$$

$$\hat{x}_{k+1} = \hat{x}_k + b'_k(y\text{-}\hat{x}_k). \tag{18}$$

If $0 < b'_k < 1$, then $\hat{x}_\infty = y$ and $\hat{x}_k \in [\bar{y}, y]$.
The performance of the IMMSE denoising technique is the function of:

- The choice of the initial filtered image $\hat{x}_0$: The initial filtered image $\hat{x}_0$ must ensure a high speckle reduction level. In [26], the boxcar filter was selected as an initial filter. However, it has been demonstrated that the use of a more sophisticated filter ensured better filtering performance [20–25].

- The choice of the parameter b'_k. In fact, this parameter is a tuning factor that controls the performance of the filtering process as the parameter b in (15). To ensure robust speckle denoising, this parameter must satisfy three important properties:

  i. $0 \leq b'_{\mathrm{k}} \leq 1$.

  ii. $b'_k \approx 0$ in homogeneous areas,

  iii. $b'_k \approx 1$ in heterogeneous areas.

Hence, by implementing N iterations (N is sufficiently low), the denoising procedure maintained the filtered homogeneous areas (i. e.$\hat{x}_N \approx \hat{x}_0$ since $b'_k \approx 0$) and preserved spatial details (i. e.$\hat{x}_N = y$ since $b'_k \approx 1$).

By the analogy of the MMSE expression of the parameter b (15), the parameter b' has been expressed as

$$b'_k = \frac{\mathrm{var}(\hat{x}_k)}{\left((1+\sigma_v^2)\mathrm{var}(\hat{x}_k) + \hat{x}_k{}^2\sigma_v^2\right)} \tag{19}$$

In [21], the authors proposed a more sophisticated version expressed as

$$b'_k = \tanh\left(\frac{\mathrm{CV}_{\hat{x}_k}^2\ \mathrm{CV}_y^2}{\mathrm{C}}\right) \tag{20}$$

$$\mathrm{C} = \frac{1}{(ENL_0)^2}, \tag{21}$$

where CV is the coefficient variation and ENL_0 is the equivalent number of looks of the original image y estimated in a homogenous area,

$$CV(y) = \frac{std(y)}{\bar{y}} \tag{22}$$

$$ENL_0 = \frac{(\bar{y})^2}{\mathrm{var}(y)} \tag{23}$$

where std is the standard deviation.

3.2 Practical implementation

For a given SAR image y

i. Compute the parameter C in (21).

ii. Compute $\hat{x}_0$ image by applying a filter ensuring high speckle reduction level.

iii. For a given pixel,

iv. From the selected pixels of the filtered image $\hat{x}_0$, compute $CV_{\hat{x}}$ [21].

v. From the selected pixel of the original image y, compute CV_y using the same process in iii.

vi. Compute b'_k using (20).

vii. Update the filtered pixel using (18).

viii. Apply the process for all pixels of the image.

Repeat iii to vii K iterations. K is a tuning parameter to control the speckle reduction and spatial detail preservation.

3.3 The INLP filter

The INLP is based on the statistics of the SAR intensity (i.e., multiplicative noise model (1) and the MMSE expression (14). In [26, 29–32], it has been demonstrated that.

$$\hat{x} = a\mathrm{var}(\hat{x}) + d \tag{24}$$

where

$$a = (y - x)/\mathrm{var}(x), \tag{25}$$

and

$$d = x. \tag{26}$$

Eq. (24) shows that the filtered pixel $\hat{x}$ is linearly related to its variance $\mathrm{var}(\hat{x})$. This rule is applied to estimate the INLP-filtered pixel (i.e., the parameter d or the noise-free pixel x). In the extended homogeneous area, the MMSE filtered pixel is $\hat{x} = \overline{x} \approx x$ while in the INLP filter $\hat{x} = d \approx x$ where d is estimated using a linear regression between means and their variances and not using a simple mean (i.e., $\hat{x} = \overline{x} \approx x$) as in the original MMSE denoising technique.

3.4 Practical implementation

For each pixel of the image:

i. Define a window W having N samples.

ii. Uniformly select N_i samples from W, where $N_i = N_{min}, \dots, N$. N_{min} is the smallest number of samples.

iii. Apply the original filter to all sets of N_i samples. We obtain $\hat{X} = (\hat{x}_N, \hat{x}_{N-1}, \dots, \hat{x}_{N_{\min}})$.

iv. Repeat steps ii and iii L times to obtain sufficient samples.

v. For each pixel of the filtered image, compute the vector $\mathrm{var}\hat{X} = (\mathrm{var}(\hat{x}_M), \mathrm{var}(\hat{x}_{M-1}), \dots, \mathrm{var}(\hat{x}_1))$ using the window W.

vi. For each pixel of the image, perform a linear regression between $\mathrm{var}\hat{X} = (\mathrm{var}(\hat{x}_M), \mathrm{var}(\hat{x}_{M-1}), \dots, \mathrm{var}(\hat{x}_1)), \hat{X} = (\hat{x}_N, \hat{x}_{N-1}, \dots, \hat{x}_{N_{\min}})$ and compute the filtered value (i.e., the constant d in (24)).

4. Results

EPD-ROATo assesses the performance of the studied denoising techniques, airborne and spaceborne SAR images were used (see **Figure 1a**). For the spaceborne SAR

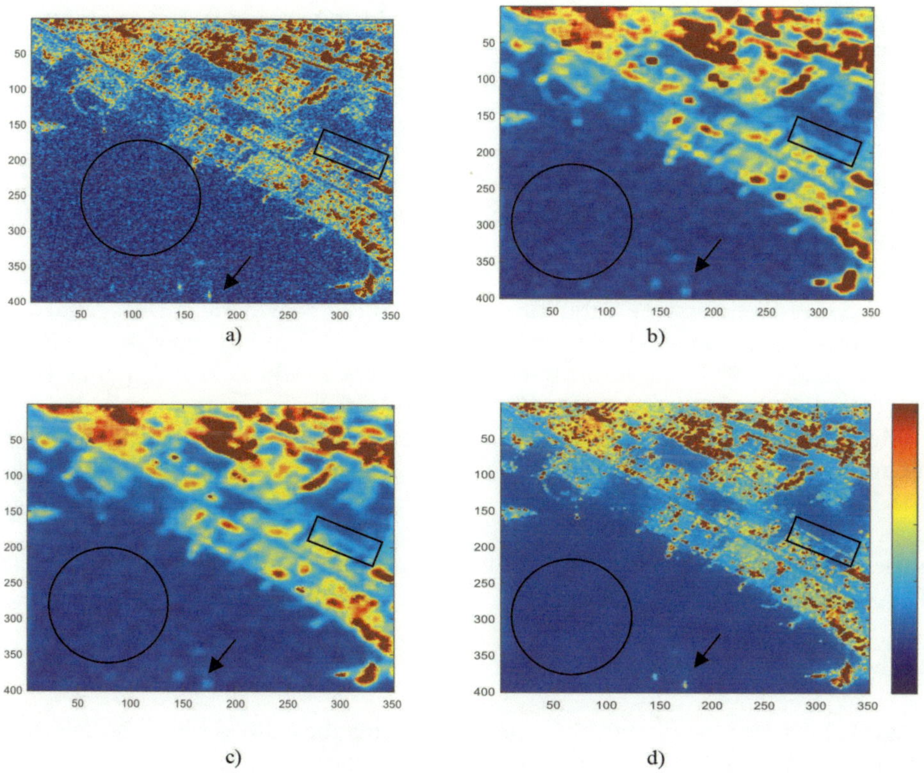

Figure 1.
(a) Original spaceborne SAR image, (b) Boxcar filter, (c) MMSE filter (i. e. improved sigma filter [18], (d) IMMSE filter [21].

data, the Sentinel 1 C-band *vv* SAR image of Dubai UAE is considered. The airborne SAR is the *hh* image of Les-Landes site, France acquired by NASA JPL AIRSAR sensor (see **Figure 2a**)).

4.1 Evaluation criteria

In addition to visual inspections, quantitative parameters have been employed to assess the performance of the studied denoising techniques. The ENL was employed to evaluate speckle reduction level

$$ENL(i) = \frac{\left(\bar{\hat{x}}(i)\right)^2}{\mathrm{var}(\hat{x}(i))}. \tag{27}$$

The edge preservation degree based on the ratio of averages (EPD-ROA) [33] is used to assess the preservation of spatial details. The EPD-ROA in horizontal direction is:

$$EPD - ROA_H(i) = \frac{\sum_{m,n} |\hat{x}(m,\ n)/\hat{x}(m,\ n+1)|}{\sum_{m,n} |y(m,\ n)/y(m,\ n+1)|}, \tag{28}$$

where m and n are the xy coordinates of the pixel in the selected zone, respectively. EPD-ROA_V is calculated by replacing in (28) the indexes (m,n + 1) by (m + 1, n). In general cases, EPD-ROA < 1. High EPD-ROA means a high ability for spatial detail preservation.

4.2 MMSE vs IMMSE

Figure 1 displays the filtered denoised spaceborne SAR images using the boxcar filter (i.e., mean filter), the MMSE (improved Lee filter), and the IMMSE filters. It can be observed that the boxcar filter reduced the speckle noise but blurred spatial details. The MMSE filter improved the filtering performance. The IMMSE maintained the high speckle reduction of the initial filter and enhanced considerably the spatial details such as lines (see rectangles) and points (see arrows). It can be seen that the IMMSE outperforms the MMSE-based filter in terms of speckle reduction and spatial detail preservation. Quantitative results in **Table 1** confirmed visual interpretations where the IMMSE filter maintained the high speckle reduction level of the initially applied filter and enhanced spatial details. Quantitative results show also that the MMSE filter outperformed the boxcar filter in terms of speckle reduction and spatial detail preservation. The IMMSE gave better filtering results than the MMSE filter in terms of speckle reduction ($ENL_{IMMSE}(1124) > ENL_{MMSE}(410)$) and spatial detail preservation ($EPD_{IMMSE}(0.98) > EPD_{MMSE}(0.94)$).

4.3 MMSE vs INLP

Figure 2 displays the filtered airborne images using the studied filters. It can be observed visually that the MMSE filter outperformed the boxcar filter. The INLP reduced the blurring effects introduced by the MMSE filter. This can be easily seen in lines (see rectangles). Concerning the speckle reduction, it is observed that the INLP filter ensured better filtering performance. In fact, the homogeneous areas appear smoother than the

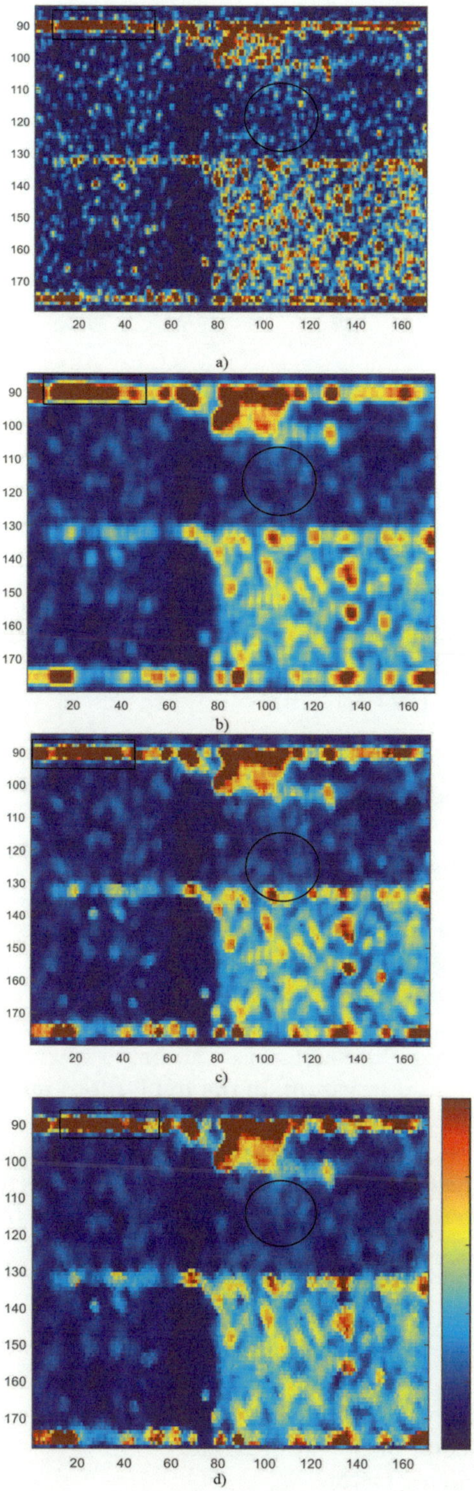

Figure 2.
(a) Original airborne SAR image, (b) Boxcar filter, (c) MMSE filter (i. e. improved sigma filter [18], (d) INLP filter [31].

	ENL	*EPD_H*	*EPD_V*
Boxcar 9×9	296	0.94	0.94
MMSE 11×11	410	0.94	0.94
IMMSE	1124	0.98	0.98

Table 1.
Performances of the filters using spaceborne data.

	ENL	*EPD_H*	*EPD_V*
Boxcar 5×5	***10***	0.83	0.83
MMSE 5×5	9	0.84	0.84
INLP	10	0.85	0.85

Table 2.
Performances of the filters using airborne data.

ones ensured by the MMSE filter (see circles). These results are recorded quantitatively in **Table 2** from which it is observed that the INLP filter outperformed the MMSE filter in terms of speckle reduction ($ENL_{INLP}(10) > ENL_{MMSE}(9)$) and spatial detail preservation ($EPD_{INLP}(0.85) > ENL_{MMSE}(0.84)$).

5. Conclusion

In this chapter, the authors reviewed the use of the MMSE-based speckle denoising techniques in SAR images. It has been shown that the MMSE-based filters (i.e., the improved Lee sigma filter) ensured high speckle denoising performance. Based on the MMSE principle, two improved MMSE versions have been introduced recently in the literature, i.e., the IMMSE and the INLP filters. The results showed that when the IMMSE is initialized with an image ensuring high speckle reduction, it ensures better denoising performance than the classical MMSE-based filters in terms of speckle filtering and spatial detail preservation. In the INLP filter, unlike the MMSE-based filters which estimated the noise-free pixels using spatial means, linear regressions between the filtered pixels and their variances for different window sizes are applied. Results show that this new strategy increased the filtering performance. Future researches will focus on the extension of the IMMSE and the INLP on additive image denoising.

Nomenclature

b	tuning factor of the MMSE filter
b'	tuning factor of the IMMSE filter
C	a normalizing factor
CV	the coefficient variation
$CV_{\hat{x}}$	the coefficient variation of $\hat{x}_k$
CV_y	the coefficient variation of y

d	the INLP filtered pixel
$E()$	mathematical expectation (i. e. statistical mean).
ENL_0	the equivalent number of looks of the original image y
EPD-ROA	the edge preservation degree based on the ratio of averages
I	mean square error
K	number of iterations.
N	the Initial number of looks
N_i	number samples selected from W
std	the standard deviation
ν	the speckle noise
W	square window
y	the intensity pixel
$\overline{y}$	spatial mean of y
x	the noise-free pixel
$\overline{x}$	spatial mean of x
$\hat{x}_0$	the initial filtered image
$\hat{x}$	filtered image
σ_ν	standard deviation of the speckle.

Author details

Mohamed Yahia[1]* and Tarig Ali[2]

1 Laboratoire de recherche modélisation analyse et commande de systèmes- MACS, Ecole Nationale d'Ingénieurs de Gabes – ENIG, Université de Gabes, Gabes, Tunisia

2 GIS and Mapping Laboratory, American University of Sharjah, Sharjah, UAE

*Address all correspondence to: mohamed_yahia1@yahoo.fr

References

[1] Lee JS, Grunes MR, De Grandi G. Polarimetric SAR speckle filtering and its implication for classification. IEEE Transations on Geoscience and Remote Sensing. 1999;**37**:2363-2373. DOI: 10.1109/36.789635

[2] Lee JS, Pottier E. Polarimetric Radar Imaging from Basics to Applications. Boca Raton, FL, USA: CRC Press; 2009. DOI: 10.1201/9781420054989

[3] Hawwar Y, Reza A. Spatially adaptive multiplicative noise image denoising technique. IEEE Transactions on Image Processing. 2002;**11**:1397-1404. DOI: 10.1109/TIP.2002.804526

[4] Argenti F, Alparone L. Speckle removal from SAR images in the undecimated wavelet domain. IEEE Transations on Geoscience and Remote Sensing. 2002;**40**:2363-2374. DOI: 10.1109/TGRS.2002.805083

[5] Vasile G, Trouvé E, Lee JS. Intensity-driven adaptive-neighborhood technique for polarimetric and interferaometric SAR parmeters estimation. IEEE Transation on Geoscience and Remote Sensing. 2006;**44**:1906-1921. DOI: 10.1109/TGRS.2005.864142

[6] Zhao Y, Liu JG, Zhang B, Hong W, Wu YR. Adaptive total variation regularization based SAR image despeckling and despeckling evaluation index. IEEE Transations on Geoscience and Remote Sensing. 2015; **53**:2765-2774. DOI: 10.1109/TGRS.2014.2364525

[7] Zhong H, Lu L, Jiao L. Fast non-local lee filter for SAR image despeckling using directional projection. Internationa l Conference on Radar. 2011. pp. 1600-1603. DOI: 10.1109/CIE-Radar.2011.6159870

[8] Deledalle CA, Denis L, Tupin F. Iterative weighted maximum likelihood denoising with probabilistic patch-based weights. IEEE Transactions on Image Processing. 2009;**18**:2661-2672. DOI: 10.1109/TIP.2009.2029593

[9] Parrilli S, Poderico M, Angelino CV, Verdoliva L. A nonlocal SAR image denoising algorithm based on LLMMSE wavelet shrinkage. IEEE Transations on Geoscience and Remote Sensing. 2012; **50**:606-616. DOI: 10.1109/TGRS.2011.2161586

[10] Penna PAA, Mascarenhas NDA. SAR speckle nonlocal filtering with statistical modeling of haar wavelet coefficients and stochastic distances. IEEE Transactions on Geoscience and Remote Sensing. 2019;**57**(9):7194-7208

[11] Foucher S, Beaulieu M, Dahmane M, Cavayas F. Deep speckle noise filtering. International Geoscience and Remote Sensing Symposium. 2019. pp. 5311-5314

[12] Foucher S, Beaulieu M, Cavayas F, Dahmane M. Revising an iterative speckle filtering technique. International Geoscience and Remote Sensing Symposium IGARSS. 2019. pp 5213-5216

[13] Denis L, Deledalle CA, Tupin F. From patches to deep learning: Combining self-similarity and neural networks for SAR image despeckling. International Geoscience and Remote Sensing Symposium IGARSS. 2019. pp. 5113-5116

[14] Lee JS. Digital image enhancement and noise filtering by use of local statistics. IEEE Transactions on Pattern Analysis and Machine Intelligence. 1980; **PAMI-2**:165-168

[15] Lee JS. Digital image smoothing and the sigma filter. Computer Vision, Graphics, and Image Processing. 1983;**24**: 255-269

[16] Kuan DT, Sawchuk AA, Strand TC, Chavel P. Adaptive noise filtering for images with signal dependent noise. IEEE Transactions on Pattern Analysis and Machine Intelligence. 1985;**PAMI-7**: 165-177

[17] Frost VS, Stiles JA, Shanmugan KS, Holtzman JC. A model for radar image and it's application to adaptive digital filtering for multiplicative noise. IEEE Transactions on Pattern Analysis and Machine Intelligence. 1982;**PAMI-4**: 175-186. DOI: 10.1109/TPAMI.1982.4767223

[18] Lee JS, Wen JH, Ainsworth TL, Chen KS, Chen AJ. Improved sigma filter for speckle filtering of SAR imagery. IEEE Transations on Geoscience and Remote Sensing. 2009;**47**:202-213. DOI: 10.1109/TGRS.2008.2002881

[19] Lee JS, Ainsworth TL, Wang Y, Chen KS. Polarimetric SAR speckle filtering and the extended sigma filter. IEEE Transations on Geoscience and Remote Sensing. 2015;**53**:1150-1160. DOI: 10.1109/TGRS.2014.2335114

[20] Fatnassi S, Yahia M, Ali T, Mortula MM. SAR speckle filtering using hybrid non-local sigma lee filter', The IEEE International - Conference on Systems, Signals and Devices. 2021

[21] Yahia M, Ali T, Mortula MM, Abdelfattah R, El Mahdi S, Arampola NS. Enhancement of SAR speckle denoising using the improved IMMSE Filter. IEEE Journal of Selected Topics in Applied Earth Observations and Remote Sensing. 2020;**13**:859-871. DOI: 10.1109/JSTARS.2020.2973920

[22] Yahia M, Ali T, Mortula MM, Abdelfattah R, El Mahdi S. Polarimetric SAR speckle reduction by hybrid iterative filtering. IEEE Access. 2020;**8**: 175-186. DOI: 10.1109/ACCESS.2020.2993902

[23] Yahia M, Ali T, Mortula MM, Abdelfattah R, El Mahdi S. Speckle denoising of the multipolarizationimges by hybrid filters. International Conference on Advanced Technologies for Signal and Image Processing. 2020

[24] Yahia M, Ali T, Mortula MM, Abdelfattah R, Arampola NS, El Mahdi S. Multitemporal SAR image despeckling based on IMMSE filtering. IEEE Mediterranean and MiddleEast Geoscience and Remote Sensing Symposium. 2020:113-116. DOI: 10.1109/M2GARSS47143.2020.9105150

[25] Fatnassi S, Yahia M, Abdelfattah R. The contribution of the non-local means to the iterative MMSE SAR despeckling. IEEE Geoscience and Remote Sensing Symposium (IGARSS). 2019

[26] Hamrouni T. Yahia M, Abdelfatah R. PolSAR speckle filtering using iterative MMSE Filter. IEEE Geoscience and Remote Sensing Symposium (IGARSS). 2018

[27] Yahia M, Ali T, Mortula M, Abdelfattah R. Effects of spatial filtering on PolSAR parameter estimation. International Conference on Advanced Technologies for Signal and Image Processing. 2022

[28] Yahia M, Ali T, Mortula M, Abdelfattah R. On MMSE speckle filtering of PolSAR images. IEEE Geoscience and Remote Sensing Symposium (IGARSS). 2022

[29] Yahia M, Hamrouni T, Abdelfatah R. SAR speckle denoising using iterative

filter. IEEE Geoscience and Remote Sensing Symposium (IGARSS). 2017

[30] Yahia M, Ali T, Mortula MM, Abdelfattah R, Arampola NS, El Mahdi S. Polarimetric SAR image filtering by infinite number of looks prediction technique. IEEE Journal of Selected Topics in Applied Earth Observations and Remote Sensing. 2021;**14**:4167-4184. DOI: 10.1109/JSTARS.2021.3070421

[31] Yahia M, Hamrouni T, Abdelfatah R. Infinite number of looks prediction in SAR filtering by linear regression. IEEE Geoscience and Remote Sensing Letters. 2017;**14**:2205-2209. DOI: 10.1109/LGRS.2017.2749322

[32] Yahia M, Aguili T. Characterization and correction of multilook effects on eigendecomposition parameters in PolSAR images. IEEE Transactions on Geoscience and Remote Sensing. 2015;**53**:5237-5246. DOI: 10.1109/TGRS.2015.2419717

[33] Feng H, Hou B, Gong M. SAR image despeckling based on local homogeneous-region segmentation by using pixel-relativity measurement. IEEE Transactions on Geoscience and Remote Sensing. 2011;**49**:2724-2737. DOI: 10.1109/TGRS.2011.2107915

Chapter 4

ANNHRPAA Based Deep Learning Image Processing for Pneumonia Detection

Avaragollada Puravarga Mathada Prasanna Kumar and S.M. Vijaya

Abstract

Pneumonia is a syndrome that is cause by a bacterial disease in the lungs. This disease is diagnosed through a chest X-ray. For triumphant treatment early diagnosis is important. This disease can be diagnosed through X-ray imagery. Sometimes due to the unclear chest X-ray image, it can be confused with the other bacterial disease. Consequently, to guide clinicians requires computer-aided diagnosis system. In this, a amalgam reverse transmission algorithms introduced by which erudition of multi-layer network achieved. The clamor investigation of the system is performed by using artificial neural network (ANN). Convolution neural network model vgg19 employed to create a user-friendly webpage for diagnosing this disease. Simulated artificial neural network hybrid reverse propagation adaptive algorithm used for deep learning image processing method in our training stage. The test results showed for the vgg19 network is at an accurateness of 0.91.

Keywords: pneumonia, transfer learning, vgg19, deep learning, webpage

1. Introduction

Mounting scientific advancement, it is potential to use tools based on unfathomable learning frameworks to discover pneumonia based on upper body X-ray imagery. The confront here would be to aid the conclusion process which allows for expedited treatment and better scientific outcome.

Pneumonia is a bacterial infection in one or both lungs which causes the inflammation of lung tissue. Over 7% of the residents which is 450 million inhabitants are affected by this disease worldwide and 4 million dies every year [1]. In India during, 2016—158,176 deaths were reported, and we continue to have the uppermost number of child deaths all over the globe. On earth pneumonia day the report was released that by 2030 over 11 million under-five children will be dead due to this transferable disease [2]. In the nineteenth century, the father of modern medicine for revolutionizing sir William Osler said pneumonia is "captain of the men of death".

The virus can easily pass from person to person which make it spread rapidly. One of the common symptoms of COVID-19 that can be easily identified is fever. Since the virus outbreak, thermal screening using infrared thermometers are used at public places to check the body temperature to identify the indicated infected among crowd. This prevention still lacking because it spends a lot of time to check the body temperature from every person and the most importance is the close contact of the infected might lead to spreading it to the person who do the screening process or from the one in charge of screening to the checked people.

Clinical examination such as chest X-ray, blood test, and other techniques are used by doctors to diagnose pneumonia in patients. In this chest X-ray is cheaper because of the technology development in bio-medical equipment. Sometimes even the clinicians fail to detect this disease by x-ray images due to the disturbance in images. Recent technology such as artificial intelligence can be useful to mitigate the disease. Especially for the image classification convolution neural network (CNNs) show great results. The main idea behind CNN is that it is an simulated model of the human brain's visual cortex. Based on the presence of pneumonia chest X-ray images are classified in convolution neural network.

2. Literature survey

The researchers [3] compared two CNN networks to diagnose pneumonia disease. To train the model they used to convey learning and fine-tuning. The consequences of the two networks are compared after the training phase. The accuracy of Xception and vgg19 are 0.82 and 0.87 respectively. And precision for Xception is 0.86 and 0.82 for the normal and pneumonia datasets. The precision for vgg16 is 0.83 and 0.91 respectively for the normal and pneumonia datasets. Here exception is more flourishing in detect pneumonia cases and vgg16 is better in detecting normal cases.

In [4] researchers tried the dissimilar technique for minimizing dimensionality. They used the JSRT dataset which has 247 X-ray images. BSE-JSRT dataset can be extracted after removing the bone shadow (dataset 02). Segmented JSRT (dataset 03) and we can have segmented BSE-JSRT (dataset 04). T-SNE technique is use to remove outlier (dataset 05). Here highest accuracy is obtained from dataset 05 which is 0.71 and the lowest accuracy is dataset 04 which is 0.56. From bone outline dataset 02 we get 0.65 accuracy.

In this paper [5], the authors used the ANN implement for detect lung diseases like pneumonia, TB. The pre- processing techniques are Lung segmentation taking out Image classification. Back-propagation and feed-forward networks are used for image classification. The dataset use from Sassoon sanatorium of 80 patients. They achieved an correctness of 0.92. The limitation is when the CXR position and size change there is no robustness. In this [6] researchers have used CNN techniques such as resnet-50 to diagnose thorax disease using chest X-ray. In pre-processing techniques, the global division take input and local branch is trained after discovering local lesion province. Here resnet-50 has average accuracy of 0.841. The AG-CNN raises the accurateness up to 0.868.

The researchers in [7, 8] created a cheXNet algorithm which as CNN of 121 layers to diagnose the pneumonia disease. They have down scaled the image to 224*224 sizes. In addition to normalization base on standard deviation and mean. The accuracy of cheX Net is 0.435. The Artificial Neural Network model by Prasanna Kumar and Vijaya [2] as Hybrid Back Propagation Adaptive Algorithm (ANNHBPAA) for clatter abolition. Adaptive clatter termination using ANN has been implementing on image signal and intelligent method for real-time signal noise cancellation based on neural networks.

Here [1, 9] the author has taken the data from 3 different hospitals for pneumonia detection. For classification, they have used the cheXNet model. And for the model training PyTorch, 0.2.0 is used. Overall, they have obtained 0.815 accuracies. But CNN does not perform well on the external data.

There has been to a great extent follow a line of investigation by Prasanna Kumar and Vijaya [10, 11] on active noise control (ANC) systems and obtainable simulated results for trans image facsimile systems. The working principle of the anticipated intelligent adaptive filter base noise cancellation system is the prolongation of prior work.

3. Proposed solutions

Known revelation intensity, the quantity of X-rays impinge on the long-suffering different at different location on the patient's remains. Confrontation of X-rays pass from side to side the patient's composition. Some are wrapped up by the patient at the same time as others exceed all the way through and are captivated by the imaging detector—an additional statistically controlled process with its own inherent noise characteristics. One time the X-rays have conceded throughout the patient, picture "information" enclosed in the spatial allocation of the X-ray fluence.

The patient's composition has shaped variation in the X-ray concentration that imaging system uses to create image. Picture "signal" is the inherent arithmetic "noise" connected with the X-ray creation method.

In distinction, when a huge quantity of radiation use, the visibility of the arithmetical noise exist very low, perhaps even hardly noticeable. Although this can outcome in a visually agreeable image, an unnecessarily high revelation level was used, consequential in overexposure to the patient.

Up to this point converse noise coupled with the statistical nature of X-ray production and their succeeding amalgamation by the patient. These process are controlled by indispensable laws of nature and, for any given X-ray acquisition, they establish the fundamental limit on image quality.

Final displayed image comes to that original threshold on image quality distinguish the total of "extra" noise that the detector introduce hooked on the image. This is fundamentally the proportion of the gesture to noise in the final image to the "unique" signal to noise at hand in the occurrence X-ray fluence. Detector forever adds some quantity of noise into the image so the DQE is forever less than 1 shown in **Figure 1A–C**.

Disperse increases as soon as imaging thicker areas of the corpse—such as the upper body. Conventional method of plummeting scatter is collimation, anti-scatter grids, and/or utilize an air-gap.

In image processing system, noise deletion using adaptive digital sieve is a well-known technique for extract most wanted images gesture by eliminate noise from the lossy picture contained indication tainted by noise. For noise annulment an assortment of gradient adaptive lattice (GAL) and LMS algorithms use. Of late, the cross adaptive algorithms with neural set of connections have gained popularity in cancelling the noise available in image compression and enhancement system. The operational principle of the planned intelligent adaptive filter-based noise cancellation system (AFNCS) is the extension of prior work Kumar et al. [12] which is additional empirically designed and computer-generated to enhance the performance of the input synthetic signal with high opinion to denoising.

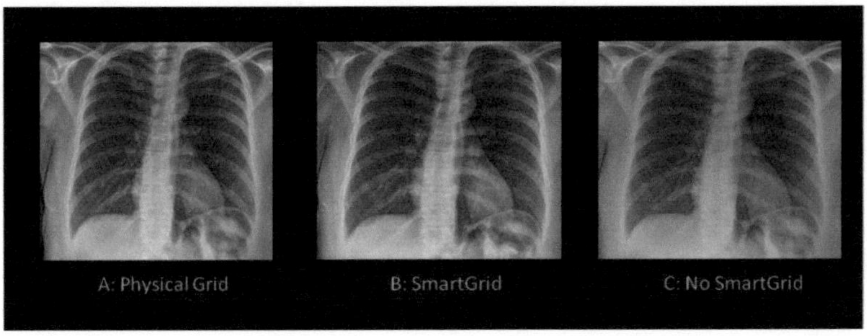

Figure 1.
A (left): Erect Portable Chest @ 105 kVp, 3.2 mAs with 6:1, 103 In/in Grid; B (center): Same patient, same SID @ 95 kVp, 2.8 mAs, no Grid, processed with Smart Grid; C (right): Same capture as B without Smart Grid.

This intelligent hybrid reverse transmission algorithm involves both GAL and LMS algorithm. The prime objective of the proposed intelligent AFNCS is to acquire signal as of reference signal and output noisy signal, in the middle of this signal noise is eliminated by subtracting the reference signal and noisy signal with original signal. Significantly reinstate the original signal by eliminate the noise by means of adaptive control and adaptation of weights from beginning to end ANN. The following **Figure 2**, indicate the chunk depiction of the AFNCS which intakes the input signal "$i(t)$" and generate signal at output "$O(t)$" by means of adaptive system and orientation signal "$R(t)$". Lastly, the signal with errore(t) is computed by finding the difference amongst reference signal and output signal as given in (1).

$$e(t) = R(t) - O(t) \tag{1}$$

Every where 't' represent number of epochs.

Implementation of mixture algorithm consider this inaccuracy signal $e(t)$ to produce a purpose for execution. This function perform the working out of required filter coefficients. The minimize error rate indicate that yield signal is similar as that of sole signal. Here reverse propagation algorithms are use to estimate the error speed of every neuron. The following **Figure 2** things to see the structural representation of

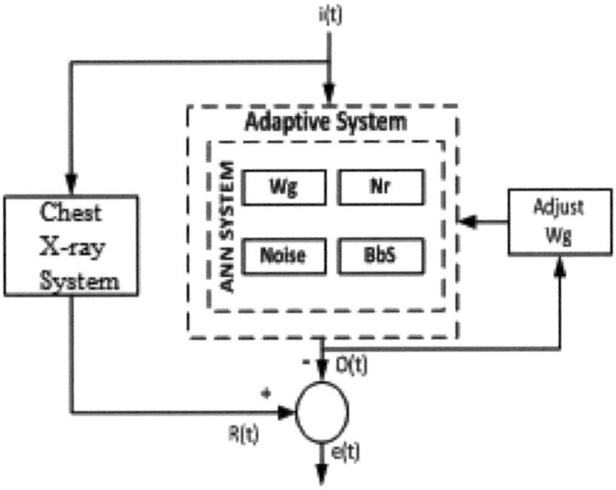

Figure 2.
Proposed adaptive filter based noise cancellation system (AFNCS).

reverse propagation level diagram of ANN network. The layer diagram of ANN network is finished up of three layers comprising input layer, concealed layer and output layer. The hidden layer is active in among input and output layer which couple both the layers. Overall back propagation network is affected by one neuron error. The network allow image signal to propagate by means of ANN and provides output signal. As given in Eq. (1) the error results of the output layer are computed and this error is forward reverse to participation layer from beginning to end hidden layer in anticipation of the considered necessary output.

Added, to reduce its inaccuracy signal, fine-tuning of weight is to execute for every neurons. Projected hybrid algorithm combine both the reverse propagation algorithm of LMS and GAL which help to embark upon sluggish convergence.

The proposed AFNCS revealed in **Figure 1** adopt adaptive filter for carrying out of ANN in addition to as well adopt a control method for fine-tuning of adaptive filter parameter. The elements association is train with ANN by credence fine-tuning. The output of ANN can be obtained by using below formula as given in (2). The following **Table 1**, indicates the parameters used in design.

$$ANN_{out} = \sum i(t) \times W_g \tag{2}$$

Each of the input are accompany by a weight.
If, $\sum W_g \geq Th$
Then the output of ANN will be 1 given in (3)

$$ANN_{out} = 1 \tag{3}$$

3.1 Data

In this study, a dataset consisting of 5842 chest X-ray images provided in **Table 1** by Guangzhou Women and Children's Medical Centre, Guangzhou. The X-ray images in the dataset are of different resolutions such as 1328 × 1160 and 1762 × 1535. The number of no pneumonia is 1576, and pneumonia is 4266. **Figure 3** shows some X-ray image samples from the dataset. In our models 0 represents normal cases, 1 represents pneumonia cases.

	Train	Test
Normal	1341	234
Pneumonia	3875	390
Total	5216	624

Table 1.
Distribution of dataset.

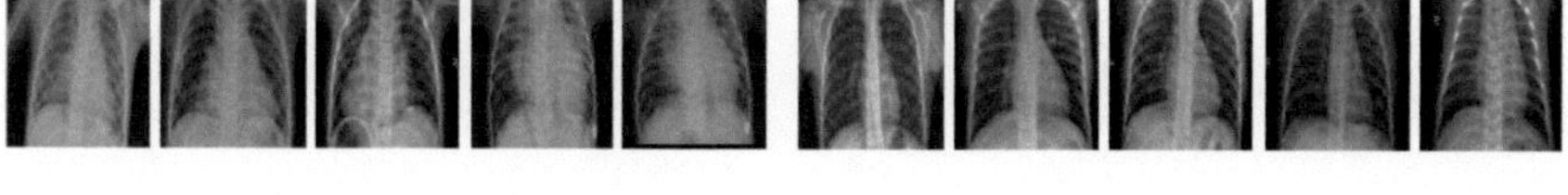

Figure 3.
Data samples from the dataset.

3.2 Pre-processing

In Deep learning, we need more data to be obtained for better and reliable results. However, there might not be more data or enough data for some problems, especially on medical problems. so, to avoid this, experts have some solutions to solve this problem. One of them is data augmentation which avoids over fitting and improves accuracy. It is supported in the Keras deep learning library image data generator class shown in **Figure 4**. Here we use rescale, shear range, Zoom range, Horizontal flip. We pre-process our X-ray images dataset before it is used for diagnosing pneumonia. The pre-processing has been performed as in following:

Unify X-ray images. Before inputting the images into our model, we downscale the images to 224 × 224 and convert them to a NumPy array. It can be suitable for features extraction by VGG. Perform image data argumentation methods, it is supported in the Keras deep learning library via the image data Generator class. Here we use rescale, shear range, Zoom range, Horizontal flip.

3.3 Architecture

AlexNet, AlexNetOWTBn, GoogleNet, VGG models are the most commonly used in transfer learning. They are a stack of many convolution layers. we have many difficulties with deep Convolution neural networks they are optimization of the network, desertion gradient problem, and deprivation problems. The VGG NET brings a new idea in place. It is used to solve complicated tasks and also increases detection accuracy. VggNet tries to resolve the difficulty in the training process of deep Convolution neural networks, the saturation, and degradation of correctness. In this paper, we have used Vgg19 architecture shown in **Figure 5**. Vgg19 network Vgg19 has 19 layers (16 convolution layers, 3 fully connected layers, 5 MaxPool layers, and 1 SoftMax layer).

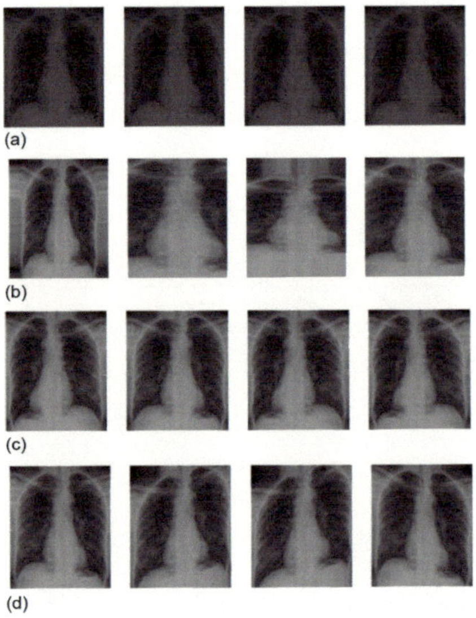

Figure 4.
(a) Rescale, (b) zoom range, (c) horizontal flip and (d) shear range for we use rescale, shear range, Zoom range, Horizontal flip. Pre-process our X-ray images dataset before it is used for diagnosing pneumonia.

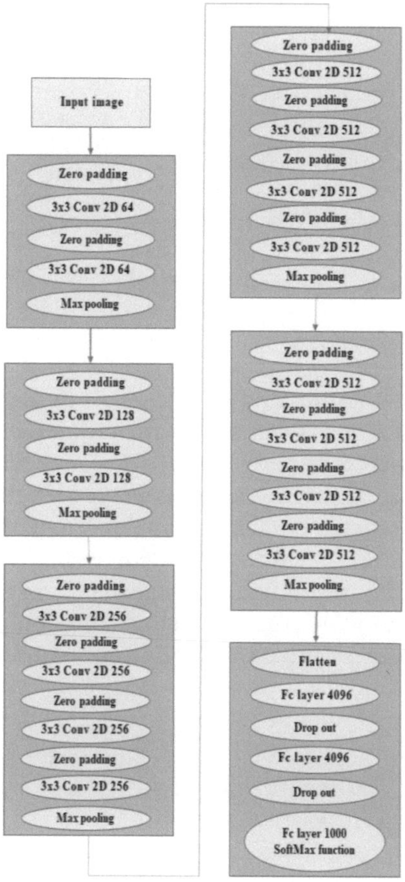

Figure 5.
Vgg19 network.

3×3 filters are used in the first and second layers in the convolutional layer. Here in the first and second layer totally 64 layers are used which results in $224 \times 224 \times 64$ volume as the same convolution used. 3×3 filters are always used with a stride of 1. The next layer is the pooling layer, here to reduce the width and height volume from $224 \times 224 \times 64$ to $112 \times 112 \times 64$ we use the max pool of 2×2 size and stride of 2 Next it is followed by 2 convolution layers which as 128 filters. Therefore, it gives the new dimension of $112 \times 112 \times 128$. Here pooling layer is used again to reduce the size to $56 \times 56 \times 128$. Now 256 filters of 2 convolution layers are added then it is reduced to $28 \times 28 \times 256$ by down sampling layer. Then the stack of 3 convolution layers is separated with 1 max-pooling layer. Finally, in the last pooling layer, we get $7 \times 7 \times 512$ volume which is flattened into a fully connected layer with a total channel of 4096 and 1 classes of soft Max output.

4. Hardware explanation

In the projected method, the convergence speed of error signal increase with the value of S_t. LMS mechanism is adopt in the proposed method because of its easier

Figure 6.
Hardware experimental board.

accomplishment, easy computational, dynamic usage of memory capability and is performed by adjusting filter coefficient for error reduction.

To estimate the performance of the projected adaptive noise cancellation algorithm by replication, the proposed algorithm is implemented on the experimental panel. As revealed in **Figure 6**, the experimental board includes one major board and one D to A/A to D data exchange card. The 16-bit D/A data exchange card is used to produce two signals. One signal is the communication signals.

Initially, the time impediment opinion performance and noise cancellation performance are evaluated in different mixed SNR environment, in that order. Secondly, the noise cancellation performance of proposed algorithm is evaluated when the time delay between the primary input and reference input is changing.

5. Experimental method

In order to authenticate the feasibility of the projected algorithm adaptive noise cancellation system based is built on FPGA, which is revealed in **Figure 7**.

In a mathematical computing atmosphere projected model by means of soft computation-based algorithm design and implementation. The system stipulation required for performance includes a 64-bit operating system, an x64-based processor supported with 4.00 GB installed memory (RAM), where the processor type is Intel® Core™ i-8250U,CPU@1.80GHz

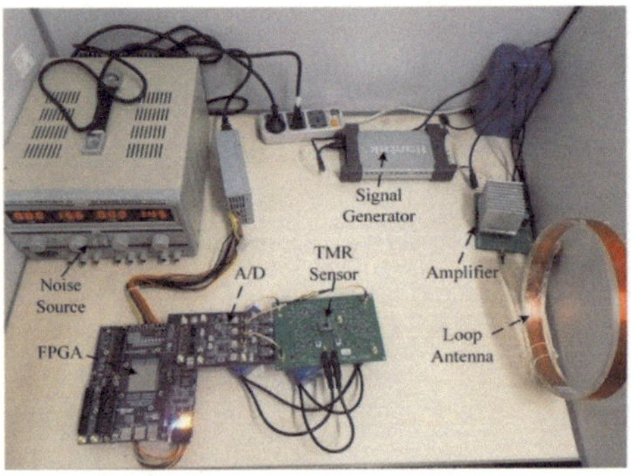

Figure 7.
Adaptive noise cancellation system.

5.1 Performed tests

We have tried many testing in different experimental setups to analyze the performance of the proposed model. We have changed several network parameters and instructions to create the model. We have split the total dataset into 80% for training purposes and 20% for validation purposes. Then, we have experimented with the dataset with our proposed model.

5.2 Fine-tuning

Fine-enhancement is a method used to increase the effectiveness of a task. It make small changes to improve the outcome. Changing the parameters is so critical that several modify affect the training process a lot for the calculation time desirable the swiftness of convergence and the use of doling out units. Parameters setup for the proposed model given in **Table 2**. This process of fine-tuning was repeated again and again to improve the accuracy of our model.

5.3 Training

We have collected 5842 X-ray images in total as our database from Guangzhou Women and Children's Medical Centre, Guangzhou, where the number of no pneumonia is 1576, and pneumonia is 4266. All the images are graded into 2 classes (NORMAL & PNEUMONIA) by professional graders and used to train the model. And it is tested with 624 images.

To train the model, we have used the pretrained vggNet, which is initialized with weights trained on ImageNet which gave better results.

5.4 Performance of the proposed model

The model which we have created will start training with the training dataset which consists of both the actual images and the images from the augmentation Then we have used the validation dataset to generalize the model.

Furthermore, we can see the spreading of losses (both training loss and validation loss) concerning the number of epochs in both the training and validation phases

In this paper with the proposed model, the X-ray images were resized into 224 × 224. Then we have done the data augmentation. We used the weights of the pretrained vgg19 model. We have used Adam optimizer, and we have used the SoftMax

Parameter	Value
Batch size	32
Steps per epoch	5216
Epoch	20
Validation steps	624
optimizer	Adam optimizer

Table 2.
Parameters setup for the proposed model.

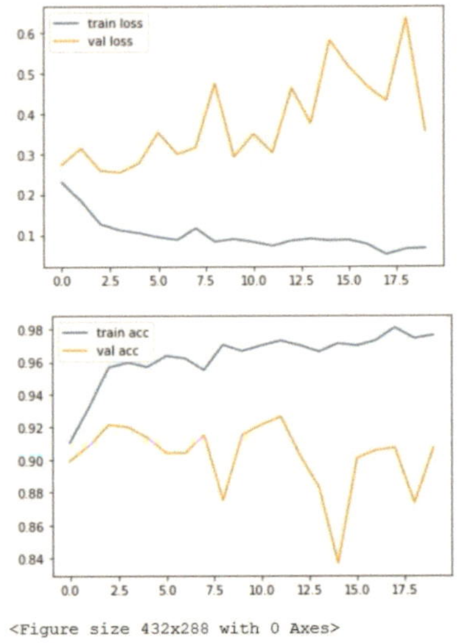

Figure 8.
Pretrained VGG-19 performance for pneumonia prediction task.

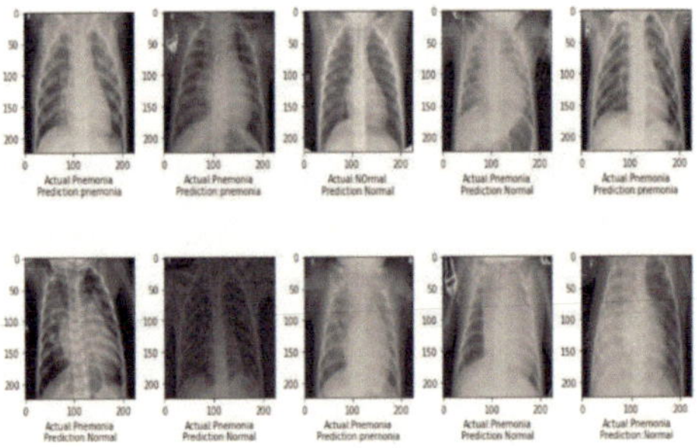

Figure 9.
Output of the model predicted with real data.

activation function and batch size equals 32. In our model, we have set the learning rate, decay, momentum as default values.

Then we started training our vgg19 model, after training, we have got the accuracy score of the model which is 0.91 where we have used the standard ImageNet weights to train the model shown in **Figures 8** and **9**.

We have trained our model up to 20 epochs; the training was stopped owing to the absence of further improvement in both accuracy and loss.

Difference between actual and predicted is given in corresponding error Column for the 6 neuron layers obtained for 5000 iterations shown in **Table 3** and in **Figure 10** gives Comparison of LMS, GAL, hybrid correlation coefficient for 5000 and 10,000 iterations.

6. Results

To predict pneumonia disease, we have created a webpage using flask API. Once Flask API is designed. We can add the trained h5 file in the flask API then we can use

Signal type	Actual LMS	Predicted	Error	Actual GAL	Predicted	Error	Hybrid algorithms	Predicted	Error
Chirp	0.8401	0.8651	0.0250	0.9218	0.9305	0.0087	0.8501	0.8663	0.0162
Sinusoidal	0.9464	0.9736	0.0272	0.9422	0.9378	−0.0044	0.9459	0.9501	0.0042
SawTooth	0.8935	0.8911	−0.0024	0.9021	0.8651	−0.0370	0.8909	0.9477	0.0568
Image signal	0.9798	0.9601	−0.0197	0.9989	0.9736	−0.0253	0.9988	0.9477	−0.0511
		Total error: 0.0301			Total error: −0.0580			Total error: 0.0261	

Table 3.
Neurons hidden layer, 5000 iterations.

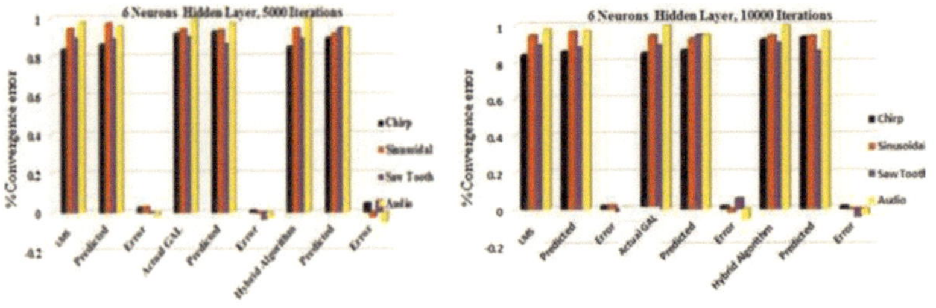

Figure 10.
Comparison of LMS, GAL, hybrid correlation coefficient for 5000 and 10,000 iterations.

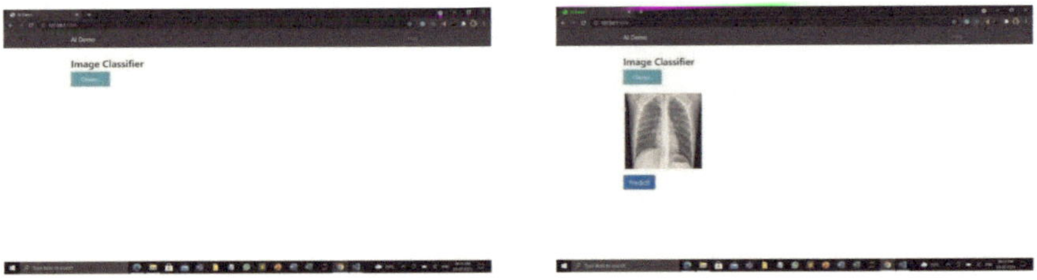

Figure 11.
Webpage which predicts the disease when input is given.

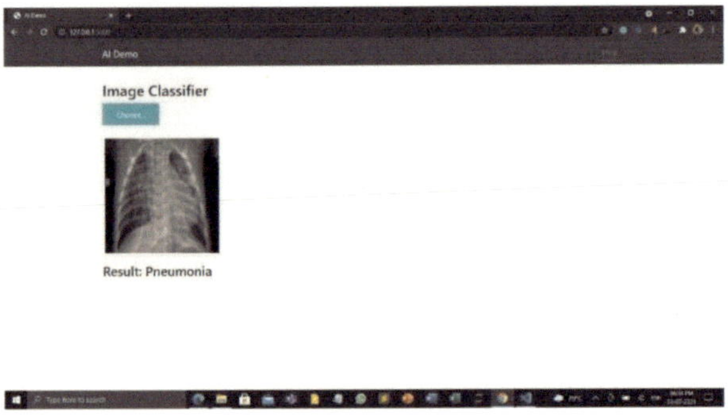

Figure 12.
Predicts the disease.

the flask run command in the command prompt to run the flask file and create a running webpage link which we can put in the browser to see the webpage.

Figure 11 shows the pneumonia disease input screen. Where user can input their X-ray image by pressing the upload button, once the user clicks on the predict button it will return whether the patient has pneumonia disease or not **Figure 12** shows the output of the predicted results.

	Algorithm	Accuracy
Base paper result	Vgg16, Xception	0.87, 0.82
Performance attainment	Vgg19	0.91

Table 4.
Performance attainment.

7. Bench mark

In the base paper they have used vgg16 and Xception model for performing training. We have used extension of vgg which is vgg19, which as more trainable parameter and gives better accuracy than vgg16 which used in our base paper. In vgg16 we have 138 million parameter and in vgg19 we have 144 million parameters. Vgg19 is the deeper version vgg16 (**Table 4**).

8. Conclusion

Projected hybrid adaptive algorithms participation signals are deterministic. LMS as well as GAL algorithms are stochastic. Adaptive noise annulment using hybrid adaptive algorithms implement. Compare in the midst of conventional algorithms, the hybrid adaptive algorithms reveal that extremely fast convergence. Amid persistent enhancement of the adaptive hybrid algorithm in addition to the rapid development of signal processing chip it will be further widely use in mobile telecommunication system, in addition to signal processing fields. The simulation perception investigation of hybrid adaptive algorithms is conceded out on the convergence behaviour, correlation coefficient and convergence time. After comparing, simulated results were tabulated. By taking into consideration of accessible algorithms performance of hybrid adaptive algorithms gives enhanced convergence time, convergence behaviour, correlation coefficients. This technique is more systematic in eliminate noise from corrupted signal furthermore has less time to converge, faster response and reduction in memory.

Convolution Neural Network used to identify the pneumonia disease automatically. To train this model employed transfer learning method and carried out fine-tuning to improve the performance of the model, our model can distinguish between 2 classes of pneumonia or normal. The Vgg19 model which we have used has shown significant performance. Results obtained confirm attained valid accuracy up to 0.91 for classifying the pneumonia disease. Inference that our model has great practical significance in early pneumonia screening and diagnosis and has strong potential to be applied in other disease.

Author details

Avaragollada Puravarga Mathada Prasanna Kumar[1*] and S.M. Vijaya[2]

1 Department of Electronics and Communication Engineering, ACS College of Engineering, Bengaluru, Karnataka, India

2 Department of Electronics and Communication Engineering, Rajarajeswari College of Engineering, Bengaluru, Karnataka, India

*Address all correspondence to: amprasannakumar@acsce.edu.in

References

[1] Zech JR, Badgeley MA, Liu M, Costa AB, Titano JJ, Oermann EK. Variable generalization performance of a deep learning model to detect pneumonia in chest radiographs: A cross-sectional study. PLoS Medicine. 2018;**15**(11):e1002683

[2] Prasanna Kumar AM, Vijaya SM. ANNHBPAA based noise cancellation employing adaptive digital filters for mobile applications. Journal of the Institution of Engineers (India). 2021; **102**(4):645-653

[3] Ayan E, Murat H. Diagnosis of Pneumonia from Chest X-Ray Images Using Deep Learning. Kırıkkale, Turkey: Kırıkkale University, IEEE; 2019

[4] Gang P, Wang Z, Zeng W, Gordienko Y, Kochura Y, Alienin O, et al. Dimensionality reduction in deep learning for chest x-ray analysis of lung cancer. In: ICACI 2018: 10th International Conference on Advanced Computational Intelligence, Xiamen, China. IEEE; 2018. pp. 878-883

[5] Khobragade S, Tiwari A, Patil CY, Narke V. Automatic detection of major lung diseases using Chest Radiographs and classification by feed-forward artificial neural network. In: IEEE 1st International Conference on Power Electronics, Intelligent Control and Energy Systems. Delhi, India: Institute of Electrical and Electronics; 2016. pp. 1-5

[6] Udeshani KAG, Meegama RGN, Fernando TGI. Statistical feature-based neural network approach for the detection of lung cancer in chest x-ray images. International Journal of Image Processing (IJIP). 2011;**5**(4):425-434

[7] Guan Q, Huang Y, Zhong Z, Zheng Z, Zheng L, Yang Y. Diagnose like a radiologist: Attention guided convolutional neural network for thorax disease classification. arXiv preprint arXiv:1801.09927 (2018)

[8] Rajpurkar P, Irvin J, Zhu K, Yang B, Mehta H, Duan T, et al. Chexnet: Radiologist-level pneumonia detection on chest x-rays with deep learning. arXiv preprint arXiv:1711.05225 (2017)

[9] Pingale TH, Patil HT. Analysis of cough sound for pneumonia detection using wavelet transform and statistical parameters. In: 2017 International Conference on Computing, Communication, Control and Automation (ICCUBEA). 2017. pp. 1-6 [Online]

[10] Prasanna Kumar AM, Vijaya SM. Noise cancellation employing adaptive digital filters for mobile applications. Indonesian Journal of Electrical Engineering and Informatics (IJEEI). 2020;**8**(1):112-121

[11] Vijaya SM, Suresh K. An efficient design approach of ROI based DWT using vedic and wallace tree multiplier on FPGA platform. International Journal of Electrical and Computer Engineering. 2019;**9**(4):24333

[12] Chan H-P, Sahiner B, Hadjiyski L, Zhou C, Petrick N. Lung nodule detection and classification. U.S. Patent Application 10/504,197, September 22, 2005

Chapter 5

Multi-Metric Near-Optimal Image Denoising

Kenji Hara and Kohei Inoue

Abstract

It is necessary to optimize the parameters for each image input to achieve the maximum denoising performance because the performance of denoising algorithms depends largely on the selection of the associated parameters. The commonly used objective image quality measures in quantitatively evaluating a denoised image are PSNR, SSIM, and MS-SSIM, which assume that the original image exists and is fully available as a reference. However, we do not have access to such reference images in many practical applications. Most existing methods for no-reference denoising parameter optimization either use the estimated noise distribution or a unique no-reference image quality evaluation measure. In the chapter, for BM3D, which is a state-of-the-art denoising algorithm, we introduce a natural image statistics (NIS) based on the generalized Gaussian distribution (GGD) and the elastic net regularization (EN) regression method and propose its use to perform the BM3D parameter optimization for PSNR, SSIM, and MS-SSIM, respectively, which are the popular image quality evaluation measures, without reference image and knowledge of the noise distribution. Experimental results with several images demonstrate the effectiveness of the proposed approach.

Keywords: denoising parameter optimization, BM3D, full-reference image quality, Kullback-Leibler divergence, elastic net regularization regression

1. Introduction

Image denoising is used for various tasks, such as segmentation, enhancement, frequency decomposition, and local feature extraction. The performance of denoising algorithms generally depends largely on the selection of the parameters. We address the problem of optimizing the parameters of denoising algorithms to achieve maximum performance. The most common image quality evaluation metrics used to quantitatively evaluate the performance of denoising methods include full-reference metrics such as peak signal-to-noise ratio (PSNR), structural similarity index measure (SSIM) [1], and multi-scale structural similarity index measure (MS-SSIM) [2], which assume that the original image exists and is fully available as a reference. However, such reference images are not available in many practical applications. Therefore, no-reference image denoising approaches have been intensively developed.

Several methods for optimizing denoising parameters without reference have been proposed that use cross-validation [3, 4] and the L-curve method [5, 6]. These methods are somewhat empirical and hence not necessarily optimization methods in the strict sense. Subsequently, a class of parameter optimization methods [7–9] was developed to minimize an estimate of the mean-squared error (MSE) obtained using Stein's unbiased risk estimate (SURE) [10]. This approach performs PSNR optimization without requiring a reference. Although PSNR is not necessarily a good evaluation measure of image quality, it remains among the most popular objective image quality metrics. However, this SURE-based approach generally necessitates the accurate estimation of the noise variance in the noisy image, which is not trivial.

Recently, Zhu et al. [11] proposed a no-reference optimization method based on a perceptual and no-reference image quality evaluation metric, which requires no knowledge of the noise distribution. Their image quality evaluation metric is very interesting in itself, but is not necessarily widely used. To the best of knowledge, most existing methods for no-reference denoising parameter optimization either use the estimated image noise or individual no-reference image quality evaluation metrics. The only exception is the no-reference parameter optimization method by D'Elia et al. [12], which requires no estimation of noise statistics and achieves its optimality only under the SSIM metric. However, PSNR and MS-SSIM metrics are also widely used for the assessment of image quality. Thus, a denoising algorithm to guarantee the near-optimality with respect to the non-SSIM criterion, particularly for PSNR and MS-SSIM, is proposed in this chapter. The proposed framework can also easily be extended to incorporate any full-reference image quality measurement metrics that might be discovered in the future.

In the chapter, we propose a novel technique for no-reference parameter optimization in the BM3D denoising algorithm, which is the current state-of-the-art denoising method. Our method adaptively depends on which of the following most widely used full-reference image quality evaluation metrics is optimized: PSNR, SSIM, and MS-SSIM, and requires no knowledge of the noise distribution. To do so, we introduce a natural image statistics (NIS) model based on the generalized Gaussian distribution (GGD) and an elastic net regularization regression model. The pipeline of our method is illustrated in **Figure 1**. Experimental results using SIDBA images are presented to show the effectiveness of the proposed method.

The reminder of chapter is organized as follows. In Section 2, we describe a statistical model of natural images in the wavelet transform domain. In Section 3, we formulate a minimization problem of a statistical distance measure to estimate the optimal parameters under the SSIM and MS-SSIM metrics. In Section 4, we derive a regression-based bias correction procedure to estimate the optimal parameter under the PSNR metric by refining a quasi-optimal parameter. In Section 5, we

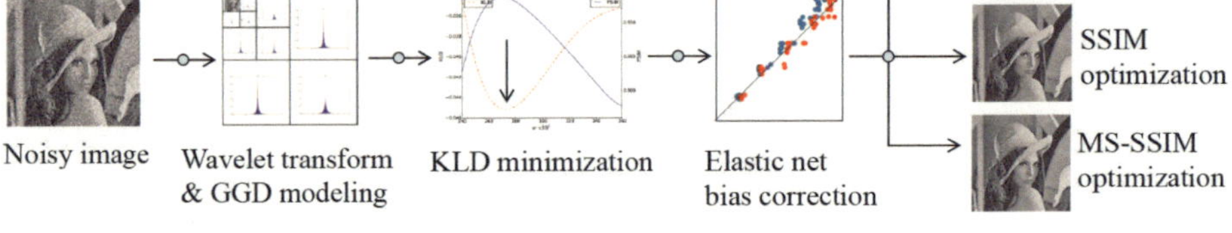

Figure 1.
Pipeline of our method.

present experimental results obtained by applying the proposed framework to each state-of-the-art denoising method. Section 5 concludes the chapter.

2. Statistical model of natural images

In this section, we use a slight modification of Mallat's statistical model of natural images [13] based on generalized Gaussian modeling in the wavelet transform domain. The generalized Gaussian distribution (GGD) [14] refers to a family of symmetric distributions, which includes the Gaussian, the Laplacian, and the uniform distributions as special cases. Recently, GGD has been successfully used in the fields of pattern recognition and image processing, in applications such as texture retrieval [15], digital watermarking [16], face recognition [17], and image segmentation [18–20].

The probability density function (pdf) of GGD with a mean zero is given by

$$p(x;\alpha,\beta) = \frac{\beta}{2\alpha\Gamma(1/\beta)} e^{-|x/\alpha|^{\beta}}, \tag{1}$$

where α and β are, respectively, the scale and shape parameters (GGD parameters) and $\Gamma(z) = \int_0^{\infty} e^{-t} t^{z-1} dt \ \ (z > 0)$ is the gamma function. For $\beta = 2$, Eq. (2) is equivalent to the Gaussian distribution, whereas for $\beta = 1$, it is equivalent to the Laplace distribution. When $\beta \to 0+$, Eq. (2) becomes a Dirac delta function distribution, and when $\beta \to +\infty$, the distribution converges to a uniform distribution, as shown in **Figure 2**, and when $\beta \to +\infty$, the distribution converges to a uniform distribution. The maximum-likelihood estimation (MLE) [21, 22], moment-based [23], and global convergence (GCM) [24] methods are widely used to estimate the GGD parameters (α, β). As described below, a statistical feature for natural image is described as a set of the estimated parameters of GGD from the marginal distribution of multiresolution wavelet coefficients of a given set of training images [25].

In the proposed approach, we first acquire a large number (K) of grayscale natural images. We first apply the discrete wavelet transformation (DWT) to each component image to decompose it into high-pass subbands (SB) of three different frequency levels and three different orientations, including horizontal subbands $\{H_j\}_{j=1,2,3}$,

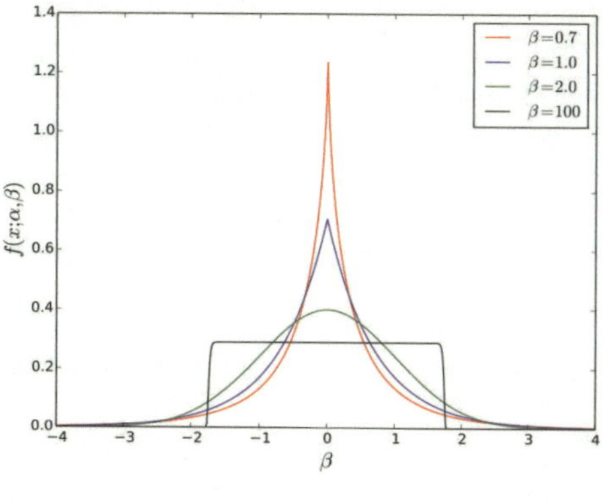

Figure 2.
The probability density functions of generalized Gaussian distribution with different shape parameters β.

vertical subbands $\{V_j\}_{j=1,2,3}$, diagonal subbands $\{D_j\}_{j=1,2,3}$, and low-pass subbands A_3, where $j = 1,2,3$ denotes the resolution level of the DWT. Then, for each of the $9(= 3 \times 3)$ high-pass subband components $\left\{\{G_j\}_{G=H,V,D}\right\}_{j=1,2,3}$, of each image, we generate a normalized histogram (SB histogram) with a bin width $h = 3.5s/n^{1/3}$, where n is the number of pixels and s is the standard deviation of the pixel values, using Scott's rule [26]. For each of the 9 SB histograms of each image, we estimate the GGD parameters $\left\{\left\{\hat{\alpha}_{G_j}^{(k)}, \hat{\beta}_{G_j}^{(k)}\right\}_{G=H,V,D}\right\}_{j=1,2,3}$, by the MLE method [21, 22], where $k = 1,\cdots,K$ is the image number. Finally, as a learning result from natural images, we obtain a set of GGDs for the 9 SB components whose pdfs are given by

$$p_{G_j}(x) = f\left(x; \overline{\alpha}_{G_j}, \overline{\beta}_{G_j}\right), \tag{2}$$

where $f(\cdot; \cdot, \cdot)$ is the pdf of GGD defined by Eq. (2). $\overline{\alpha}_{G_j}$ and $\overline{\beta}_{G_j}$ are respectively the mean values of the total number K of the estimated α and β parameters for each SB component; that is,

$$\overline{\alpha}_{G_j} = \frac{1}{K}\sum_{k=1}^{K} \hat{\alpha}_{G_j}^{(k)}, \quad \overline{\beta}_{G_j} = \frac{1}{K}\sum_{k=1}^{K} \hat{\beta}_{G_j}^{(k)}. \tag{3}$$

3. Parameter quasi-optimization

In the study, we address the problem of optimizing the parameters of the BM3D algorithm (block-matching and 3D filtering) [27]. The BM3D algorithm is designed to denoise images corrupted with zero-mean additive Gaussian noise. A modified version called SAR-BM3D has also been proposed, which assumes multiplicative speckle noise.

Firstly, the algorithm divides a noisy image into blocks. The similar blocks are stacked together to form a 3D array. Then, based on the structural similarity in each group, collaborative filtering and weighted averaging are carried out. The BM3D algorithm requires that an associated parameter σ be set to the noise variance of the image. However, accurate estimation of the noise distribution present in an image is not trivial. In addition, such an optimal denoising parameter generally varies depending on the selection of evaluation criteria.

Here, we consider that the quasi-optimal parameter $\overline{\sigma}$ is obtained as the value of σ selected such that a statistical difference between the set of SB histograms of the denoised image using BM3D algorithm and the set of GGD pdfs is minimized. We adopt as the Kullback-Leibler divergence (KLD) [28] as one of the most widely used statistical measures. We solve the following optimization problem.

$$\overline{\sigma} = \underset{\sigma}{argmin} \sum_{j=1}^{3} \sum_{G \in \{H,\ V,\ D\}} D_{KL}\left(P_{G_j} \,\|\, Q_{G_j}(\sigma)\right), \tag{4}$$

where $D_{KL}\left(P_{G_j} \,\|\, Q_{G_j}(\Theta)\right)$ denotes the KLD between two distributions P_{G_j} and $Q_{G_j}(\Theta)$ as

$$D_{KL}\left(P_{G_j} \| Q_{G_j}(\sigma)\right) = \sum_i p_{G_j}(x_i) \log \frac{p_{G_j}(x_i)}{Q_{G_j}(i|\sigma)}, \quad (5)$$

where x_i and $Q_{G_j}(i|\sigma)$ $(G = H, V, D; j = 1,2,3)$ are the center and the value of the i-th bin in each SB histogram of the denoised image using BM3D algorithm, respectively. $p_{G_j}(\cdot)$ $(G = H, V, D; j = 1,2,3)$ is the GGD pdfs defined in Eq. (3).

However, the quasi-optimal parameter $\overline{\sigma}$ did not necessarily achieve an optimal denoising in some evaluation criteria. Thus, we used a paired t test at a significance level of $\alpha = 0.05$, to determine whether there was a statistically significant difference between the quasi-optimal parameter $\overline{\sigma}$ and each of σ^*_{PSNR}, σ^*_{SSIM}, and $\sigma^*_{\text{MS-SSIM}}$, which denote the ground-truth optimal parameters for the popular image quality evaluation metrics PSNR, SSIM, and MS-SSIM, respectively. As described in Section 5, our experimental results suggest that there is a statistically significant difference in only PSNR between the quasi-optimal and ground-truth optimal parameters. Thus, hereinafter we assume that both of the estimated SSIM-optimal parameter $\hat{\sigma}_{\text{SSIM}}$ and the estimated MS-SSIM-optimal one $\hat{\sigma}_{\text{MS-SSIM}}$ are given by the quasi-optimal parameter $\overline{\sigma}$ and that there was a bias between the ground-truth PSNR-optimal parameter σ^*_{PSNR} and $\overline{\sigma}$. In the next section, we describe a method to correct the bias to obtain the estimated PSNR-optimal parameter $\hat{\sigma}_{\text{PSNR}}$.

4. Regression-based bias correction

We generate N training pairs of noisy and noise-free images by adding zero-mean Gaussian noise of different levels of noise variance and different random seed numbers to original images. Let $(y_i)_{\text{PSNR}}$ $(i = 1, \cdots, N)$ (the subscript PSNR is omitted henceforth for brevity) be the objective variable that is the ground-truth PSNR-optimal parameter σ^*_{PSNR} for the i-th training pair. Let $\boldsymbol{x}_i = \left(x_i, x_i^2, \cdots, x_i^p\right)^{\text{T}}$ $(i = 1, \cdots, N)$ be the explanatory variable vector, where x_i is the estimated quasi-optimal parameter $\overline{\sigma}$ from the i-th noisy image. Let $(\xi_0, \xi)_{\text{PSNR}}$ be the regression parameter, where $\xi_0 \in \mathbb{R}$ and $\boldsymbol{\xi} = \left(\xi_1, \xi_2, \cdots, \xi_p\right)^{\text{T}} \in \mathbb{R}^p$.

Ordinary least squares regression is commonly used to perform polynomial regression. Least squares regression is a simple method, but it is widely known that a more stable and interpretable solution is obtained by incorporating regularization into the solution of ordinary least squares. In such regularization regression models, lasso regression is a typical and well-known approach to impose a L1 norm penalty [29]. However, if there are training samples with high correlation as the noisy training images in our learning system, lasso tends to select only one sample and ignore others. Therefore, the bias correction described in this section is achieved with an elastic net [30], which is a robust regression model and avoids this problem. Using the elastic net regularization regression, the regression parameter is obtained by solving the following optimization problem.

$$\left\{\hat{\xi}_0, \hat{\boldsymbol{\xi}}\right\} = \underset{(\xi_0, \boldsymbol{\xi}) \in \mathbb{R}^{p+1}}{\operatorname{argmin}} \sum_{i=1}^{N} \left(y_i - \xi_0 - \boldsymbol{x}_i^T \boldsymbol{\xi}\right)^2 + P_\lambda(\boldsymbol{\xi}), \quad (6)$$

where the regularization term $P_\lambda(\boldsymbol{\xi})$ is expressed as a linear combination of the L1 norm $\|\boldsymbol{\xi}\|_1$ and the L2 norm $\|\boldsymbol{\xi}\|_2^2$ as

$$P_\lambda(\boldsymbol{\xi}) = \lambda_1\|\boldsymbol{\xi}\|_1 + \lambda_2\|\boldsymbol{\xi}\|_2^2 = \lambda_1\sum_{j=1}^{p}|\xi_j| + \lambda_2\sum_{j=1}^{p}\xi_j^2, \tag{7}$$

where λ_1 and λ_2 are the positive magnitudes of the L1 and L2 norm penalties, respectively. By using the solutions of Eq. (7), the estimated PSNR-optimal parameter $\hat{\sigma}_{PSNR}$ is expressed as follows.

$$\hat{\sigma}_{\mathrm{PSNR}} = \overline{\sigma} + \sum_{j=0}^{p}\left(\hat{\xi}_j\right)_{\mathrm{PSNR}}\overline{\sigma}^j. \tag{8}$$

Note that for correctness the subscript PSNR is shown explicitly in Eq. (8). In our experiments, we used the degree p of the polynomial as $p = 3$ and the tuning parameter λ of the elastic net as $\lambda_1 = \lambda_2 = 5$.

5. Experimental results

$K = 6000$ training images described in Section 3 were randomly selected from photography websites such as http://pro.foto.com and http://sozaing.com. All the images were cropped to be the size of 256×256 pixels. **Figure 3(a, b, c)** shows the plots of the KLD (red curves, left vertical axes) and PSNR, SSIM and MS-SSIM (blue curves, right vertical axes) for different values of the BM3D parameter σ (horizontal axes) for the Lena image corrupted by Gaussian noise $N(0, 30)$, respectively. From **Figure 3**, it may be observed that the quasi-optimal parameter $\overline{\sigma}$, which is the value of σ minimizing KLD, was slight smaller than the ground-truth PSNR-optimal parameter σ^*_{PSNR}, whereas $\overline{\sigma}$ was almost equal to the ground-truth SSIM-optimal and MS-SSIM-optimal parameters σ^*_{SSIM} and $\sigma^*_{\mathrm{MS-SSIM}}$.

Next, a total of $50\,(= 5 \times 5 \times 2)$ noisy SIDBA images (Lena, Pepper, Airplane, Parrots, and Girl) corrupted by five different zero-mean Gaussian noise $N(0, \sigma_n), (\sigma_n = 10, 20, \cdots, 50)$ for different two seeds of random number generator were prepared as training images, and then, the regularization regression described in Section 4 was applied on the training image set to determine the relationships between the quasi-optimal parameter $\overline{\sigma}$ and each of the ground-truth optimal parameters σ^*_{PSNR}, σ^*_{SSIM}, and $\sigma^*_{\mathrm{MS-SSIM}}$.

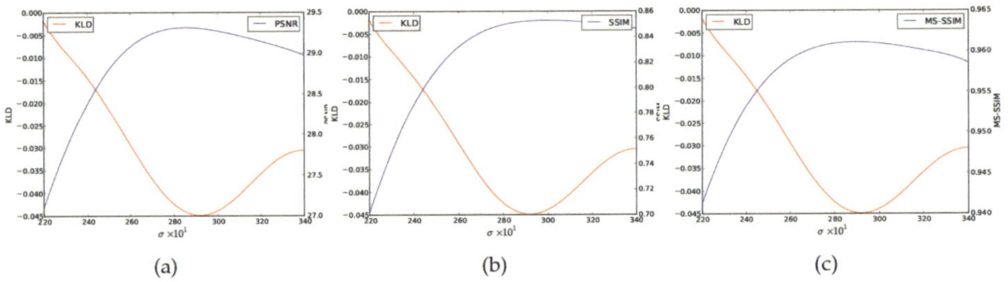

Figure 3.
Plots of KLD (red curves) and popular image quality evaluation metrics (blue curves) versus different BM3D parameters σ. (a) KLD and PSNR. (b) KLD and SSIM. (c) KLD and MS-SSIM.

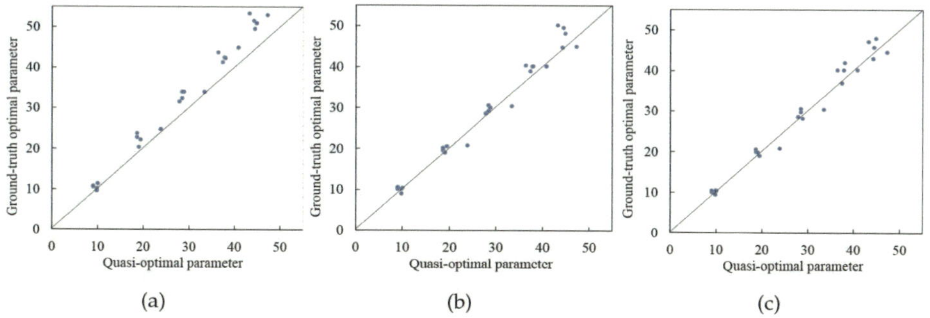

Figure 4.
Plots of ground-truth optimal parameters versus different quasi-optimal parameters $\overline{\sigma}$. *(a)* σ^*_{PSNR} *versus* $\overline{\sigma}$. *(b)* σ^*_{SSIM} *versus* $\overline{\sigma}$. *(c)* $\sigma^*_{\mathrm{MS-SSIM}}$ *versus* $\overline{\sigma}$.

To this end, first we qualitatively evaluated the denoising parameter quasi-optimization described in Section 3. **Figure 4(a, b, c)** shows the plots of the ground-truth optimal parameters σ^*_{PSNR}, σ^*_{SSIM}, and $\sigma^*_{\mathrm{MS-SSIM}}$ (vertical axes) versus the quasi-optimal parameter $\overline{\sigma}$ (horizontal axes), respectively. From **Figure 4(a)**, it may be observed that $\overline{\sigma}$ is smaller than σ^*_{PSNR} and the trend becomes more significant for larger noise variance. In contrast, from **Figure 4(b, c)**, it may be observed that $\overline{\sigma}$ was very close to both of σ^*_{SSIM} and $\sigma^*_{\mathrm{MS-SSIM}}$. Furthermore, to quantitatively evaluate the denoising parameter quasi-optimization, we used a paired t test at a significance level of $\alpha = 0.05$ to compare the quasi-optimal parameter $\overline{\sigma}$ with each of the ground-truth optimal parameters σ^*_{PSNR}, σ^*_{SSIM}, and $\sigma^*_{\mathrm{MS-SSIM}}$. The results showed no statistically significant differences for SSIM (the paired t test yielded a p value of $0.1167 > 0.05$) and MS-SSIM (the paired t test gives a p value of $0.1744 > 0.05$), and hence, we estimate the SSIM-optimal and MS-SSIM-optimal parameters as $\hat{\sigma}_{\mathrm{SSIM}} = \hat{\sigma}_{\mathrm{MS-SSIM}} = \overline{\sigma}$. In contrast, the above results showed a statistically significant difference for PSNR (the paired t test gives a p value of $5.2962 \times 10^{-8} < 0.05$), and thus, we calculated Eq. (8) to find the estimated PSNR-optimal parameter $\hat{\sigma}_{\mathrm{PSNR}}$.

To qualitatively evaluate the parameter optimization described in Section 4, we illustrate in **Figure 5(a)** the plots of the ground-truth PSNR-optimal parameter σ^*_{PSNR} (vertical axis) versus the quasi-optimal parameter $\overline{\sigma}$ (blue plots, horizontal axis) and the estimated PSNR-optimal parameter $\hat{\sigma}_{\mathrm{PSNR}}$ (red plots, horizontal axis) for the

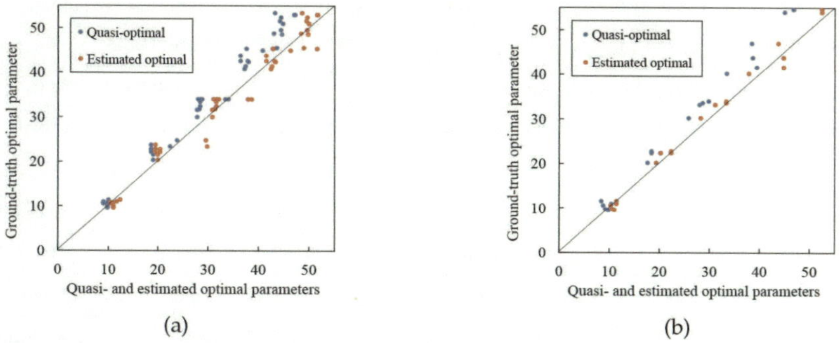

Figure 5.
Plots of ground-truth PSNR-optimal parameters σ^*_{PSNR} *versus different quasi-optimal parameters* $\overline{\sigma}$ *(blue points) and estimated PSNR-optimal parameters* $\hat{\sigma}_{\mathrm{PSNR}}$ *(red points) on the training and test image sets. (a) The training image set. (b) The test image set.*

training image set. From **Figure 5(a)**, it may be observed that bias correction based on the elastic net regularization regression improved the quasi-optimal parameters. To quantitatively evaluate the parameter optimization described in Section 4, we used a paired t test at a significance level of $\alpha = 0.05$ to compare the estimated PSNR-optimal parameter with the ground-truth PSNR-optimal parameter. The results showed no statistically significant differences (the paired t test gives a p value of $p = 0.2692 > 0.05$).

In **Tables 1** and **2**, we summarize the pairs of the ground-truth optimal parameters and the metric values and the pairs of the estimated optimal parameters and the

Image	σ_n	Ground-truth	Estimated	Ground-truth	Estimated	Ground-truth	Estimated
		(σ^*_{PSNR}, PSNR)	($\hat{\sigma}_{PSNR}$, PSNR)	(σ^*_{SSIM}, SSIM)	($\hat{\sigma}_{SSIM}$, SSIM)	($\sigma^*_{MS-SSIM}$, MS-SSIM)	($\hat{\sigma}_{MS-SSIM}$, MS-SSIM)
Lena	10	(10.8, 37.49)	(10.4, 37.48)	(10.2, 0.9950)	(9.0, 0.9946)	(10.0, 0.9930)	(9.0, 0.9927)
	20	(22.8, 35.42)	(19.5, 35.29)	(19.6, 0.9887)	(18.6, 0.9884)	(20.0, 0.9847)	(18.6, 0.9843)
	30	(32.4, 34.35)	(31.6, 34.31)	(29.2, 0.9814)	(28.4, 0.9814)	(29.8, 0.9755)	(28.4, 0.9752)
	40	(42.6, 33.52)	(43.1, 33.52)	(40.2, 0.9722)	(37.8, 0.9715)	(40.2, 0.9644)	(37.8, 0.9611)
	50	(49.6, 32.79)	(49.6, 32.78)	(49.6, 0.9643)	(44.4, 0.9626)	(45.8, 0.9531)	(44.4, 0.9527)
Pepper	10	(11.4, 35.89)	(10.4, 35.88)	(10.4, 0.9930)	(10.0, 0.9930)	(10.4, 0.9917)	(10.0, 0.9916)
	20	(24.8, 34.19)	(25.7, 34.17)	(20.8, 0.9848)	(23.8, 0.9843)	(20.8, 0.9837)	(23.8, 0.9833)
	30	(34.0, 33.46)	(37.8, 33.46)	(30.4, 0.9754)	(33.4, 0.9749)	(30.4, 0.9744)	(33.4, 0.9738)
	40	(45.0, 32.90)	(46.2, 32.89)	(40.2, 0.9638)	(40.8, 0.9636)	(40.2, 0.9631)	(40.8, 0.9630)
	50	(53.0, 32.41)	(51.6, 32.40)	(45.0, 0.9525)	(47.2, 0.9522)	(44.6, 0.9522)	(47.2, 0.9517)
Airplane	10	(10.2, 36.53)	(10.9, 36.53)	(10.2, 0.9635)	(9.8, 0.9634)	(9.4, 0.9903)	(9.8, 0.9902)
	20	(22.2, 35.21)	(20.3, 35.17)	(20.6, 0.9371)	(19.4, 0.9358)	(19.0, 0.9787)	(19.4, 0.9786)
	30	(34.0, 34.51)	(32.0, 34.49)	(30.0, 0.9137)	(28.8, 0.9125)	(28.2, 0.9666)	(28.8, 0.9663)
	40	(41.4, 33.68)	(40.6, 33.66)	(39.0, 0.8893)	(37.4, 0.8884)	(37.0, 0.9529)	(37.4, 0.9529)
	50	(51.6, 32.78)	(50.4, 32.76)	(44.8, 0.8720)	(44.2, 0.8717)	(43.0, 0.9424)	(44.2, 0.9417)
Parrots	10	(10.6, 37.74)	(10.4, 37.73)	(10.6, 0.9889)	(9.0, 0.9873)	(10.4, 0.9905)	(9.0, 0.9897)
	20	(23.8, 35.46)	(20.4, 35.41)	(20.2, 0.9770)	(18.6, 0.9753)	(20.6, 0.9793)	(18.6, 0.9780)
	30	(34.0, 34.46)	(31.5, 34.43)	(30.6, 0.9645)	(28.4, 0.9631)	(30.6, 0.9675)	(28.4, 0.9662)
	40	(42.4, 33.88)	(43.3, 33.86)	(40.2, 0.9512)	(38.0, 0.9462)	(42.0, 0.9549)	(38.0, 0.9475)
	50	(51.0, 33.32)	(50.9, 33.31)	(48.2, 0.9379)	(44.8, 0.9362)	(48.0, 0.9420)	(44.8, 0.9400)
Girl	10	(9.6, 35.53)	(10.9, 35.43)	(9.0, 0.9385)	(9.8, 0.9373)	(10.4, 0.9878)	(9.8, 0.9877)
	20	(20.4, 33.95)	(19.9, 33.94)	(19.0, 0.9044)	(19.0, 0.9044)	(19.8, 0.9742)	(19.0, 0.9741)
	30	(31.6, 33.22)	(30.8, 33.19)	(28.6, 0.8711)	(27.8, 0.8706)	(28.6, 0.9569)	(27.8, 0.9566)
	40	(43.8, 32.74)	(41.4, 32.72)	(40.4, 0.8384)	(36.4, 0.8323)	(40.2, 0.9376)	(36.4, 0.9353)
	50	(53.4, 32.33)	(48.5, 32.32)	(50.2, 0.8066)	(43.2, 0.8046)	(47.2, 0.9179)	(43.2, 0.9177)

Table 1.
Comparison of the ground-truth optimal parameters, the ground-truth metric values, the estimated optimal parameters, and the estimated metric values. The comparisons are shown for the training image set.

Image	σ_n	Ground-truth	Estimated	Ground-truth	Estimated	Ground-truth	Estimated
		(σ^*_{PSNR}, PSNR)	($\hat{\sigma}_{\text{PSNR}}$, PSNR)	(σ^*_{SSIM}, SSIM)	($\hat{\sigma}_{\text{SSIM}}$, SSIM)	($\sigma^*_{\text{MS-SSIM}}$, MS-SSIM)	($\hat{\sigma}_{\text{MS-SSIM}}$, MS-SSIM)
Balloon	10	(10.6, 38.64)	(10.26, 38.61)	(10.0, 0.9767)	(8.8, 0.9736)	(10.2, 0.9888)	(8.8, 0.9876)
	20	(22.4, 36.54)	(20.2, 36.44)	(19.6, 0.9515)	(18.4, 0.9489)	(20.6, 0.9721)	(18.4, 0.9709)
	30	(33.2, 35.52)	(31.1, 35.46)	(30.2, 0.9287)	(28.0, 0.9254)	(29.6, 0.9533)	(28.0, 0.9520)
	40	(47.0, 34.90)	(43.7, 34.87)	(40.2, 0.9068)	(38.4, 0.8960)	(40.2, 0.9348)	(38.4, 0.9235)
	50	(56.4, 34.36)	(51.7, 34.34)	(48.4, 0.8886)	(45.8, 0.8860)	(48.4, 0.9167)	(45.8, 0.9155)
Couple	10	(9.8, 36.33)	(10.5, 36.29)	(9.2, 0.9303)	(9.2, 0.9303)	(9.6, 0.9872)	(9.2, 0.9871)
	20	(20.2, 33.60)	(19.3, 33.59)	(18.8, 0.8535)	(17.6, 0.8530)	(17.6, 0.9665)	(17.6, 0.9665)
	30	(30.2, 32.66)	(28.2, 32.63)	(27.4, 0.7847)	(25.8, 0.7835)	(26.0, 0.9420)	(25.8, 0.9419)
	40	(40.2, 32.25)	(37.8, 32.22)	(35.6, 0.7216)	(33.4, 0.7199)	(34.0, 0.9157)	(33.4, 0.9154)
	50	(56.8, 32.18)	(49.3, 32.14)	(43.4, 0.6704)	(43.0, 0.6702)	(43.0, 0.8893)	(43.0, 0.8893)
Earth	10	(11.0, 36.26)	(11.4, 36.24)	(10.8, 0.9862)	(10.4, 0.9862)	(10.2, 0.9899)	(10.4, 0.9897)
	20	(22.4, 34.57)	(22.3, 34.57)	(20.6, 0.9711)	(20.2, 0.9709)	(17.4, 0.9759)	(20.2, 0.9742)
	30	(34.0, 33.74)	(33.3, 33.74)	(31.0, 0.9551)	(29.8, 0.9549)	(27.0, 0.9582)	(29.8, 0.9557)
	40	(41.6, 33.11)	(44.8, 33.09)	(40.2, 0.9405)	(39.4, 0.9351)	(34.0, 0.9347)	(39.4, 0.9295)
	50	(54.6, 32.55)	(52.4, 32.53)	(47.2, 0.9257)	(46.8, 0.9256)	(43.0, 0.9191)	(46.8, 0.9141)
Barbara	10	(11.6, 37.40)	(11.4, 37.38)	(10.8, 0.9702)	(8.4, 0.9599)	(10.8, 0.9951)	(8.4, 0.9940)
	20	(22.8, 35.26)	(22.3, 35.25)	(21.0, 0.9396)	(18.4, 0.9324)	(21.4, 0.9872)	(18.4, 0.9862)
	30	(33.6, 34.24)	(33.3, 34.23)	(30.4, 0.9029)	(28.6, 0.9756)	(30.4, 0.9761)	(28.6, 0.9557)
	40	(43.8, 33.81)	(44.8, 33.79)	(40.2, 0.8717)	(38.6, 0.8462)	(40.2, 0.9662)	(38.6, 0.9578)
	50	(54.0, 33.33)	(52.4, 33.32)	(46.8, 0.8374)	(45.0, 0.8348)	(45.8, 0.9532)	(45.0, 0.9530)

Table 2.
Comparison of the ground-truth optimal parameters, the ground-truth metric values, the estimated optimal parameters, and the estimated metric values. The comparisons are shown for the test image set.

metric values for different levels of zero-mean Gaussian noise variance and different metrics for the training image set and a test image set consisting of four SIDBA images (Balloon, Couple, Earth, and Barbara; hereinafter are referred to as "the testing image set"), respectively.

In **Figure 5(b)**, we show the plots of the ground-truth PSNR-optimal parameter (vertical axis) versus the quasi-optimal parameter (blue plots, horizontal axis) and the estimated PSNR-optimal parameter (red plots, horizontal axis) for the test image set. From **Figure 5(b)**, as well as the results in **Figure 5(a)**, our bias correction visually improves the quasi-optimal parameters. As in **Figure 5(a)**, we used a paired t test at a significance level of $\alpha = 0.05$ to compare the estimated PSNR-optimal parameter with the ground-truth PSNR-optimal parameter. The results showed no statistically significant differences (the paired t test gives a p value of $p = 0.1920 > 0.05$), and hence, we can confirm the validity of $\hat{\sigma}_{\text{PSNR}}$.

Figure 6 illustrates a visual comparison of the denoising results. **Figure 6(a)** shows the standard Barbara image. **Figure 6(b)** shows the enlarged detail in the eye

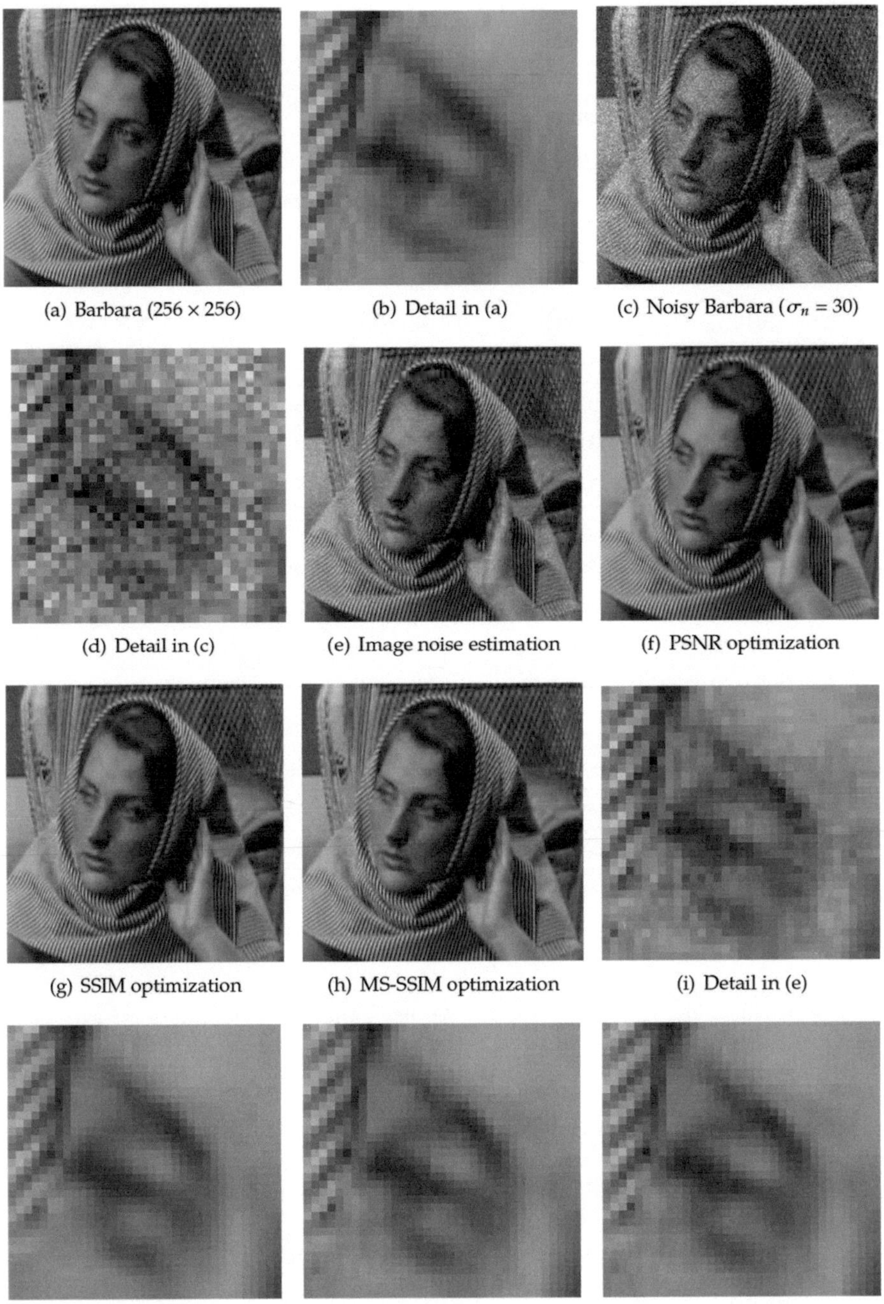

(a) Barbara (256 × 256) (b) Detail in (a) (c) Noisy Barbara (σ_n = 30)

(d) Detail in (c) (e) Image noise estimation (f) PSNR optimization

(g) SSIM optimization (h) MS-SSIM optimization (i) Detail in (e)

(j) Detail in (f) (k) Detail in (g) (l) Detail in (h)

Figure 6.
Visual comparison of denoising results. (a) Barbara image. (b) Enlarged detail of (a). (c) Image corrupted by Gaussian noise $N(0,\sigma_n = 30)$. (d) Enlarged detail of (c). (e) Denoised image with the BM3D algorithm the parameter σ of which is the estimated noise variance from (c) ($\hat{\sigma}_n = 20.6$). (f) Denoised image with the BM3D algorithm the parameter σ of which is the estimated PSNR-optimal parameter $\hat{\sigma}_{PSNR} = 33.3$. (g) Denoised image with the BM3D algorithm the parameter σ of which is the estimated SSIM-optimal parameter $\hat{\sigma}_{SSIM} = 28.6$. (h) Denoised image with the BM3D algorithm the parameter σ of which is the estimated MS-SSIM-optimal parameter $\hat{\sigma}_{MS-SSIM} = 28.6$. (i) Enlarged detail of (e). (j) Enlarged detail of (f). (k) Enlarged detail of (g). (l) Enlarged detail of (h).

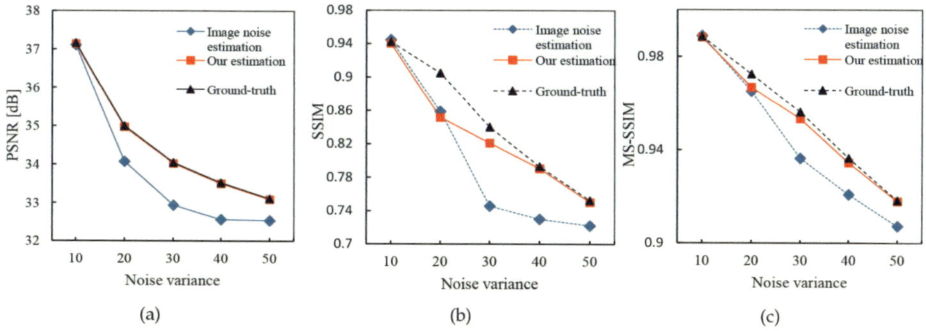

Figure 7.
Plots of image quality evaluation metric values versus different levels of image noise variance for the direct estimation of the image noise variance (blue lines and points), our proposed method (red lines and points), and the ground-truth (black lines and points). The comparisons are shown for the test image set. (a) PSNR. (b) SSIM. (c) MS-SSIM.

area of the original Barbara image shown in (a). **Figure 6(c)** shows the noisy Barbara image corrupted by additive zero-mean Gaussian noise $N(0, \sigma_n = 30)$. **Figure 6(d)** shows an enlarged detail in (c). **Figure 6(e)** shows the denoised image by the BM3D with the directly estimated noise variance from the noisy Barbara image shown in (c) as the parameter value. **Figure 6(f, g, h)** shows the denoised images by the BM3D with the estimated PSNR-optimal, SSIM-optimal and MS-SSIM-optimal parameters, respectively. **Figure 6(i–l)** shows the enlarged details in (e), (f), (g), and (h), respectively. These results indicate that the BM3D model was able to denoise the noisy image with reasonable accuracy by using our estimated optimal parameters.

Finally, for each of the image quality evaluation metrics PSNR, SSIM, and MS-SSIM, we compare the measure value of the BM3D denoised image using our estimated optimal parameters, against that using the estimated noise variance from the input noisy image. **Figure 7(a, b, c)** shows the plots of the mean values of PSNR, SSIM, and MS-SSIM at each noise variance across the test image set, respectively. From these results, it may be observed that the BM3D algorithm with our estimated optimal parameter outperformed that with the directly estimated noise variance from the input images.

6. Conclusions

We addressed the problem of estimating the optimal parameter of state-of-the-art denoising algorithm BM3D algorithm without any reference and without any knowledge of the noise distribution, adaptively depending on which of the following widely used image quality evaluation metrics are optimized: PSNR, SSIM, and MS-SSIM. The proposed method for SSIM and MS-SSIM optimization is formulated as a minimization problem for a Kullback-Leibler divergence measure based on the natural image statistics and generalized Gaussian distribution based prior. The method for PSNR optimization is formulated as a combination of the above optimization and an elastic net regression, which provides a very robust regression model. From our experimental results, we have confirmed that the proposed statistical measure and robust regression approach can be used to optimize the denoising parameter of the BM3D algorithm.

Author details

Kenji Hara*† and Kohei Inoue†
Department of Media Design, Kyushu University, Fukuoka, Japan

*Address all correspondence to: hara@design.kyushu-u.ac.jp

† These authors contributed equally.

References

[1] Wang Z, Bovik AC, Sheikh H, Simoncelli EP. Image quality assessment: From error visibility to structural similarity. IEEE Transactions on Image Processing. 2004;**13**(4):600-612. DOI: 10.1109/TIP.2003.819861

[2] Wang Z, Simoncelli EP, Bovik AC. Multi-scale structural similarity for image quality assessment. In: Proceedings of the 37th IEEE Asilomar Conference on Signals, Systems & Computers; 9-12 November 2003. Pacific Grove, CA, USA: IEEE; 2003. pp. 1398-1402

[3] Golub GH, Heath M, Wahba G. Generalized cross-validation as a method for choosing a good ridge parameter. Technometrics. 1979;**21**(2):215-223. DOI: 10.1080/00401706.1979.10489751

[4] Girard DA. The fast Monte Carlo cross validation and C_L procedures: Comments, new results and application to image recovery problems. Computational Statistics. 1995;**10**: 205-258

[5] Hansen PC. Analysis of discrete ill-posed problems by means of the L-curve. SIAM Review. 1992;**34**(4):561-580. DOI: 10.1137/1034115

[6] Hansen PC, O'Leary DP. The use of the L-curve in the regularization of discrete ill-posed problems. SIAM Journal on Scientific Computing. 1993; **14**(6):1487-1503. DOI: 10.1137/0914086

[7] Blu T, Luisieri F. The sure-let approach to image denoising. IEEE Transactions on Image Processing. 2007; **16**(11):2778-2786. DOI: 10.1109/TIP.2007.906002

[8] Luisier F, Blu T, Unser M. A new SURE approach to image denoising: Interscale orthonormal wavelet thresholding. IEEE Transactions on Image Processing. 2007;**16**(3):593-606. DOI: 10.1109/TIP.2007.891064

[9] Ramani S, Blu T, Unser M. Monte-Carlo SURE: A black-box optimization of regularization parameters for general denoising algorithms. IEEE Transactions on Image Processing. 2008;**17**(9): 1540-1554. DOI: 10.1109/TIP.2008.2001404

[10] Stein C. Estimation of the mean of a multivariate normal distribution. Annals of Statistics. 1981;**9**(6):1135-1151. DOI: 10.1214/AOS/1176345632

[11] Zhu X, Milanfar P. Automatic parameter selection for denoising algorithms using a no-reference measure of image content. IEEE Transactions on Image Processing. 2010;**19**(12): 3116-3132. DOI: 10.1109/TIP.2010.2052820

[12] D'Elia M, De Los Reyes JC, Miniguano-Trujillo A. Bilevel parameter learning for nonlocal image denoising models. Journal of Mathematical Imaging and Vision. 2021;**63**(6):753-775. DOI: 10.1007/s10851-021-01026-2

[13] Mallat SG. A theory for multiresolution signal decomposition: The wavelet representation. IEEE Transactions on Pattern Analysis and Machine Intelligence. 1989;**11**(7): 674-693. DOI: 10.1109/34.192463

[14] Box GEP, Tiao GC. Bayesian Inference in Statistical Analysis. Chichester: John Wiley & Sons; 1997. DOI: 10.1002/9781118033197

[15] Do MN, Vetterli M. Wavelet-based texture retrieval using generalized

Gaussian density and Kullback-Leibler distance. IEEE Transactions on Image Processing. 2002;**11**(2):146-158. DOI: 10.1109/83.982822

[16] Zhu X, Milanfar P. Automatic parameter selection for denoising algorithms using a no-reference measure of image content. IEEE Transactions on Image Processing. 2010;**19**(12): 3116-3132. DOI: 10.1109/ TIP.2010.2052820

[17] Jimenez DG, Gonzalez FP, Alfaro PC, Freire LP, Castro JLA. Modeling Gabor coefficients via generalized Gaussian distributions for face recognition. In: Proceedings of IEEE International Conference on Image Processing (ICIP2007); 16-19 September 2007. San Antonio, TX, USA: IEEE; 2007. pp. IV-485-IV-488

[18] Allili MS, Bouguila N, Ziou D. Finite generalized Gaussian mixture modeling and applications to image and video foreground segmentation. In: Proceedings of the IEEE Fourth Canadian Conference on Computer and Robot Vision (CRV 07); 28-30 May 2007. Montreal, Canada: IEEE; 2007. pp. 183-190

[19] Bazi Y, Bruzzone L, Melgani F. Image thresholding based on the EM algorithm and the generalized Gaussian distribution. Pattern Recognition. 2007; **40**(2):619-634. DOI: 10.1016/j. patcog.2006.05.006

[20] Fan S-KS, Lin Y. A fast estimation method for the generalized Gaussian mixture distribution on complex images. Computer Vision and Image Understanding. 2009;**113**(7):839-853. DOI: 10.1016/j.cviu.2009.03.003

[21] Joshi RL, Fischer TR. Comparison of generalized Gaussian and Laplacian modeling in DCT image coding. IEEE Signal Processing Letters. 1995;**2**(5): 81-82. DOI: 10.1109/97.386283

[22] Varanasi MK, Aazhang B. Parametric generalized Gaussian density estimation. The Journal of the Acoustical Society of America. 1998;**86**(4):1404-1415. DOI: 10.1121/1.398700

[23] Yu S, Zhang A, Li H. A review of estimating the shape parameter of generalized Gaussian distribution. The Journal of Computer Information Systems. 2012;**8**(21):9055-9064

[24] Song KS. A globally convergent and consistent method for estimating the shape parameter of a generalized Gaussian distribution. IEEE Transactions on Information Theory. 2006;**52**(2): 510-527. DOI: 10.1109/TIT.2005.860423

[25] Morinaga A, Hara K, Inoue K, Urahama K. Classification between natural and graphics images based on generalized Gaussian distributions. Information Processing Letters. 2018; **138**:31-34. DOI: 10.1016/j.ipl.2018. 05.010

[26] Scott DW. On optimal and data-based histograms. Biometrika. 1979; **66**(3):605-610. DOI: 10.2307/2335182

[27] Dabov K, Foi A, Katkovnik V, Egiazarian K. Image denoising by sparse 3D transform-domain collaborative filtering. IEEE Transactions on Image Processing. 2007;**16**(8):2080-2095. DOI: 10.1109/TIP.2007.901238

[28] Kullback S, Leibler RA. On information and sufficiency. The Annals of Mathematical Statistics. 1951;**22**(1): 79-86. DOI: 10.1214/aoms/1177729694

[29] Wright J, Yang AY, Ganesh AJ, Sastry SS, Ma Y. Robust face recognition via sparse representation. IEEE Transactions on Pattern Analysis and

Machine Intelligence. 2009;**31**(2): 210-227. DOI: 10.1109/TPAMI.2008.79

[30] Zou H, Hastie T. Regularization and variable selection via the elastic net. Journal of the Royal Statistical Society Series B. 2005;**67**(5):301-320. DOI: 10.1111/j.1467-9868.2005.00527.x

Chapter 6

Perspective Chapter: New Image Denoising Approach Based on SWT and 2-D Dual-Tree Discrete Wavelet Transform

Mourad Talbi and Riadh Baazaoui

Abstract

In this chapter, we propose a new image denoising approach. It consists in applying a Stationary Wavelet Transform (SWT) based image denoising technique, in the domain of 2-D Dual-Tree Discrete Wavelet Transform. In fact, this proposed approach consists first of applying the 2-D Dual-Tree Discrete Wavelet Transform to the noisy image. Then, the obtained noisy wavelet coefficients are denoised by applying to each of them a SWT based image denoising technique. Finally, the denoised image is reconstructed by applying the inverse of the 2-D Dual-Tree Discrete Wavelet Transform to the obtained denoised wavelet coefficients. For applying this SWT based image denoising technique, we use soft thresholding, the Daubechies 4 as the mother wavelet and the decomposion level is equal to 5. The performance of this proposed image denoising approach, is pouved by the results obtained from the computations of PSNR (Peak Signal-to-Noise Ratio) and SSIM (Structural Similarity).

Keywords: image denoising, 2-D dual-tree discrete wavelet transform, SWT-2D, PSNR, SSIM, standard deviation

1. Introduction

Noisy images frequently arise in the high-level vision tasks and this makes image denoising becoming an important task in the low-level vision domain [1]. For example, take a given denoising model:

$$y = x + n \tag{1}$$

With x, y and n are respectively the clean images, the given noisy image and the Additive Gaussian Noise (AWGN) having σ as standard deviation [1]. There are diverse approaches for reducing noise that various researchers have done. Each of these approaches owns its advantages and disadvantages. In [2], a review of some significant work in the domain of image denoising based on the denoising techniques were presented. These techniques can be classified as wavelet domain, spatial domain, or both techniques can combine for obtaining an advantage of them. In the medical domain, Mittal et al. [3] have presented a methodology for improving and eliminating the high noise of the medical image employing the Stationary Wavelet Transform (SWT) technique. In their work, an efficient and simple approach for adaptive noise elimination was used, the SWT-2D denoising method on the medical image that is degraded by noise. In this chapter, we propose a novel approach of Image denoising. It consists in applying a Stationary Wavelet Transform (SWT) based image denoising technique [4] in the domain of 2-D Dual-Tree Discrete Wavelet Transform. This SWT-based image denoising technique [4] is based on soft thresholding of the noisy wavelet coefficients obtained from the noisy image decomposition using the SWT-2D. For this decomposition, we use Daubechies 4 as the mother wavelet and the level is equal to 5. Those choices are the same as those in ref. [4].

This proposed denoising approach is completely different from the other denoising technique based on thresholding in the domain of 2-D Dual-Tree Complex Wavelet Transform [5]. In fact, this difference lies in the fact that we apply SWT-based image denoising technique [4] not to the noisy image to be denoised, but to each noisy wavelet coefficient obtained from the application of this Transform [5] to this image. Consequently, this idea can introduce more adaptability compared to the application of this technique based on thresholding in the domain of 2-D Dual-Tree Discrete Wavelet Transform [5], to this image.

The remaining of this chapter is organized as follows: in Section 2, we will deal with 2-D Dual-Tree Complex Wavelet Transform [5]. In Section 3, we will deal with the SWT-2D-based image denoising technique [4]. In Section 4, we will detail the image-denoising approach proposed in this work. In Section 5, we will present results and discussion and we will conclude in Section 6.

2. 2-D dual-tree complex wavelet transform

It turns out that, for some applications of *DWT* (Discrete Wavelet Transform), ameliorations can be obtained by employing an expansive wavelet transform in place of a critically sampled one [5]. An expansive transform is one that permits to convert an N-point signal into M coefficients with $M > N$. There are numerous sorts of expansive DWTs; here is described the dual-tree complex *DWT* [5, 6]. The dual-tree complex *DWT* of a signal x is implemented employing two critically sampled DWTs in parallel on the same data, as illustrated in **Figure 1**.

The transform is 2-times expansive because for an N-point signal, it permits to have $2N$*DWT* coefficients. If the filters in the upper and lower DWTs are the same, consequently no advantage is added. Though, when the filters are designed in a specific way, consequently the sub-band signals of the upper *DWT* can be considered as the real part of a complex wavelet transform, and sub-band signals of the lower *DWT*can be viewed as the imaginary part. Equivalently, for specially designed filters sets, the wavelet associated with the upper *DWT* can be viewed as an approximate Hilbert transform of the wavelet associated with the lower DWT. When designed in this manner, the dual-tree complex *DWT* is approximately shift-invariant, in contrast with the critically sampled DWT. Furthermore, the dual-tree complex DWT can be employed for implementing 2-D wavelet transforms where each wavelet is oriented, which is precisely useful for image processing. For the separable 2–D DWT, recall that

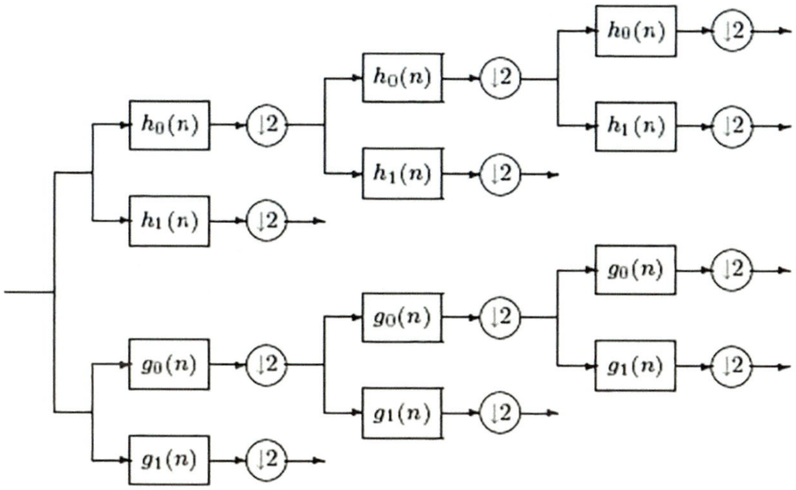

Figure 1.
The dual-tree complex DWT.

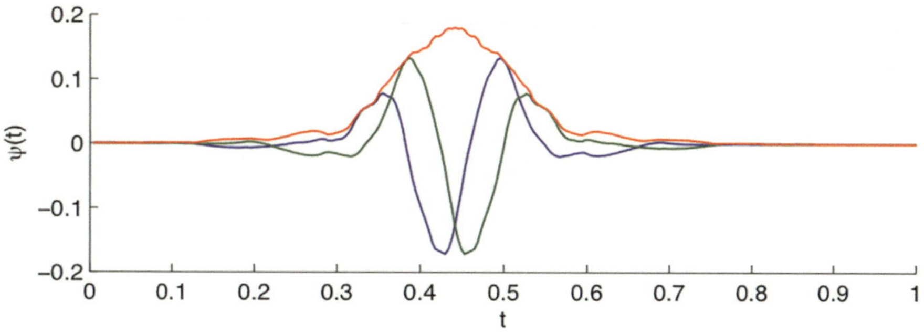

Figure 2.
Complex 1-D wavelet, $\psi(t)$ [5, 6].

one of the three wavelets does not own a dominant orientation. The dual tree complex DWT outperforms the critically sampled *DWT* for applications such as image denoising and enhancement. The complex wavelet associated with the dual-tree complex DWT is illustrated in **Figure 2**. For determining the real part of the complex wavelet, we set all coefficients to zero, except for one coefficient in the upper DWT, and after that applying the inverse transform. For determining the imaginary part, the exception is a single coefficient in the lower DWT.

3. A stationary wavelet transform (SWT) based image denoising technique

The SWT (Stationary Wavelet Transform) [7, 8] is similar to the Discrete Wavelet Transform (DWT) except the signal is never sub-sampled and instead the filters are up sampled at each level of decomposition [7]. Each level's filters are up-sampled versions of the previous as shown in **Figure 3**.

The SWT is an inherent redundant scheme, as each set of coefficients contains the same number of samples as the input. So for a decomposition of N levels, there is a redundancy of $2N$.

An image denoising technique based on thresholding in the SWT domain, is applied in our denoising system proposed in this work. This technique can be summarized by the block diagram illustrated at **Figure 4**.

According to **Figure 4**, the Stationary Wavelet Transform 2D (SWT–2D) is firstly applied to the noisy image, I_b, in order to obtain noisy stationary wavelet coefficients. Those obtained coefficients are then denoised by employing soft thresholding and finally the inverse of SWT–2D, $SWT^{-1} - 2D$ is applied to the obtained thresholded coefficients for having the denoised image, I_d.

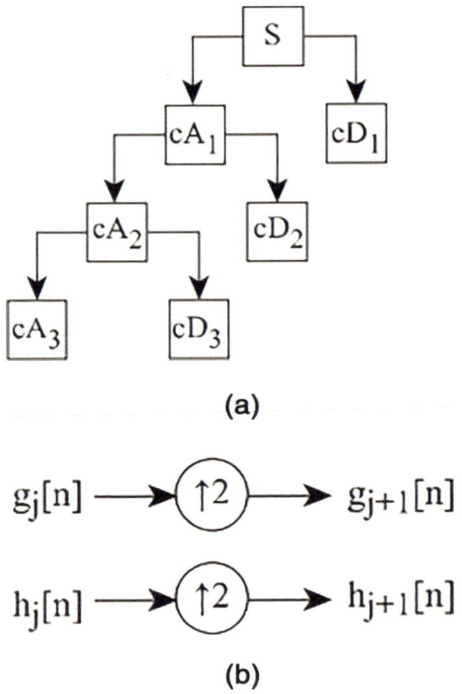

Figure 3.
(a) Wavelet decomposition tree, (b) SWT filters [7].

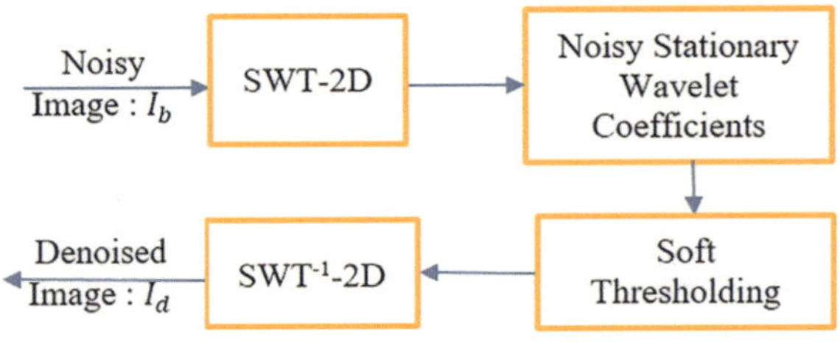

Figure 4.
The block diagram of an image denoising technique based on thresholding in the SWT domain.

4. The proposed image denoising approach

As previously mentioned, in this work, we propose a new image denoising approach. It consists in applying a SWT based image denoising technique [4] in the domain of 2-D Dual-Tree Discrete Wavelet Transform [5]. In fact, this proposed approach consists firstly in applying the 2-D Dual-Tree Discrete Wavelet Transform to the noisy image and the obtained noisy wavelet coefficients are then denoised by applying to each of them, a SWT based image denoising technique [4]. Finally, the denoised image is reconstructed by applying the inverse of the 2-D Dual-Tree Discrete Wavelet Transform to the obtained denoised wavelet coefficients. This proposed image denoising approach can be summarized by the block diagram illustrated at **Figure 5**.

According to **Figure 5**, the 2-D dual-Tree Discrete Wavelet Transform is firstly applied to the noisy image, I_b in order to obtain noisy wavelet coefficients, $Wb\{j\}\{s\}, 1 \leq j \leq 2, 1 \leq s \leq 3$. Each of those coefficients is then denoised by applying a technique of image denoising based on thresholding in SWT domain [4] and we obtain denoised wavelet coefficients, $Wd\{j\}\{s\}, 1 \leq j \leq 2, 1 \leq s \leq 3$. To those denoised coefficients, is applied the inverse of the 2-D dual-Tree Discrete Wavelet Transform in order to have finally the denoised I_d.

As previously mentioned, for the application of the denoising technique based on thresholding in the SWT domain, is used Daubechies 4 as the mother wavelet. Moreover, the decomposition level is equal to 5. Those choices are the same to those in [4]. Those choices are justified by the fact that in our comparative study, we want to use the same mother wavelet and the same decomposition level such as used in [4].

5. Results and discussion

In this section, we will present the results obtained from the computations of Peak Signal to Noise Ratio (PSNR) and Structural Similarity (SSIM). Those results are obtained from the application of the proposed image denoising approach, the denoising

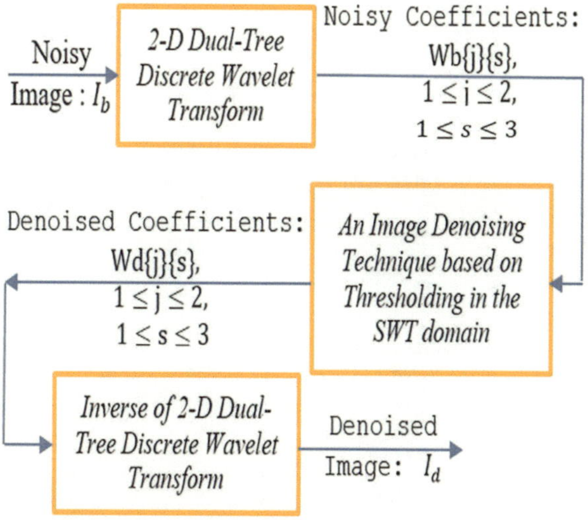

Figure 5.
The block diagram of the proposed image denoising approach.

technique based on thresholding in the SWT domain [4], the image denoising technique based on thresholding in the domain of 2-D Dual-Tree Discrete Wavelet Transform [5] and the image denoising approach using deep neural network [9].

In the following sub-section we will present the previously mentioned evaluation criterion, which are the PSNR and the SSIM. The PSNR is a better test since it takes the signal strength into consideration (not only the error). The PSNR and SSIM are expressed as follow [10]:

$$SSIM = \frac{\left(2\mu_x\mu_y + c_1\right)\left(2\sigma_{xy} + c_2\right)}{\left(\mu_x^2 + \mu_y^2 + c_1\right)\left(\sigma_x^2 + \sigma_y^2 + c_2\right)} \tag{2}$$

Where σ, σ_{xy} and μ are respectively the variance, the covariance of the image and c_1 and c_2 are the stabilizing constants. The SSIM value is generally between 0 and 1 and similar images have value of SSIM near to 1.

$$PSNR = 10 \bullet log_{10}\left(\frac{MAX_I^2}{MSE}\right) \tag{3}$$

With MSE is the Mean Square Error, expressed as follow:

$$MSE = \frac{1}{n}\sum_{i=1}^{n}\left(X_i - X_i^*\right)^2 \tag{4}$$

The MSE is one of the earliest tests which performed to test whether two images are similar.

As previously mentioned, those results (**Table 1**) are in terms of PSNR and SSIM and they are obtained for different images and diverse values of level (σ) of noise

Noisy image	**The denoising technique**			
	The proposed Image denoising technique	**The image denoising technique based on thresholding in the SWT domain [4]**	**The Image denoising technique based on deep neural network [9]**	**The image denoising technique based on thresholding in the domain of 2-D Dual-Tree Discrete Wavelet Transform [5]**
Noisy st.tif ($\sigma = 10$)	PSNR: **34.2517** SSIM: 0.8791	PSNR: 33.7831 SSIM: **0.8962**	PSNR: 28.3030 SSIM: 0.5425	PSNR: 34.1453 SSIM: 0.8675
Noisy st.tif ($\sigma = 20$)	PSNR: **30.6126** SSIM: **0.7722**	PSNR: 30.1884 SSIM: 0.6950	PSNR: 22.4725 SSIM: 0.2850	PSNR: 30.2840 SSIM: 0.7306
Noisy st.tif ($\sigma = 30$)	PSNR: **28.1244** SSIM: **0.6131**	PSNR: 24.5706 SSIM: 0.3719	PSNR: 18.9533 SSIM: 0.1789	PSNR: 27.7900 SSIM: 0.5977

Noisy image	The denoising technique			
	The proposed Image denoising technique	**The image denoising technique based on thresholding in the SWT domain [4]**	**The Image denoising technique based on deep neural network [9]**	**The image denoising technique based on thresholding in the domain of 2-D Dual-Tree Discrete Wavelet Transform [5]**
Noisy st.tif ($\sigma = 40$)	PSNR: **26.4861** SSIM: **0.5138**	PSNR: 20.6093 SSIM: 0.2171	PSNR: 16.4559 SSIM: 0.1248	PSNR: 25.9906 SSIM: 0.4953
Noisy Peppers..tif ($\sigma = 10$)	PSNR: **32.8048** SSIM: **0.8250**	PSNR: 32.4622 SSIM: 0.8019	PSNR: 28.6273 SSIM: 0.6497	PSNR: 32.6519 SSIM: 0.8180
Noisy Peppers.tif ($\sigma = 20$)	PSNR: **29.6501** SSIM: **0.7233**	PSNR: 29.3674 SSIM: 0.6821	PSNR: 22.4648 SSIM: 0.3569	PSNR: 29.5645 SSIM: 0.7038
Noisy Peppers.tif ($\sigma = 30$)	PSNR: **27.8670** SSIM: **0.6406**	PSNR: 27.8481 SSIM: 0.6264	PSNR: 18.8973 SSIM: 0.2208	PSNR: 27.4151 SSIM: 0.5968
Noisy Peppers.tif ($\sigma = 40$)	PSNR: **26.3256** SSIM: **0.5462**	PSNR: 26.2747 SSIM: 0.5424	PSNR: 16.3708 SSIM: 0.1497	PSNR: 25.7770 SSIM: 0.5109
Noisy House. tif ($\sigma = 10$)	PSNR: **35.0397** SSIM: **0.9203**	PSNR: 34.2284 SSIM: 0.8466	PSNR: 28.4578 SSIM: 0.5576	PSNR: 35.0360 SSIM: 0.9121
Noisy House. tif ($\sigma = 20$)	PSNR: **30.7882** SSIM: **0.8462**	PSNR: 30.3082 SSIM: 0.7360	PSNR: 22.4921 SSIM: 0.3090	PSNR: 30.5116 SSIM: 0.7766
Noisy House. tif ($\sigma = 30$)	PSNR: **28.1921** SSIM: **0.6586**	PSNR: 27.0232 SSIM: 0.5584	PSNR: 18.9670 SSIM: 0.2049	PSNR: 27.9388 SSIM: 0.6487
Noisy House. tif ($\sigma = 40$)	PSNR: **26.2186** SSIM: 0.5368	PSNR: 25.4000 SSIM: 0.4792	PSNR: 16.4148 SSIM: 0.1455	PSNR: 26.0214 SSIM: **0.5416**

Table 1.
Results obtained in terms of PSNR and SSIM.

corrupting the original image (clean image). This noise is an Additive Gaussian White Noise (AGWN).

In **Figure 6**, are illustrated some examples of image denoising by applying the denoising approach proposed in this work and the other previously mentioned techniques, used in our evaluation [4, 5, 9]. According to **Figure 6**, the noise is

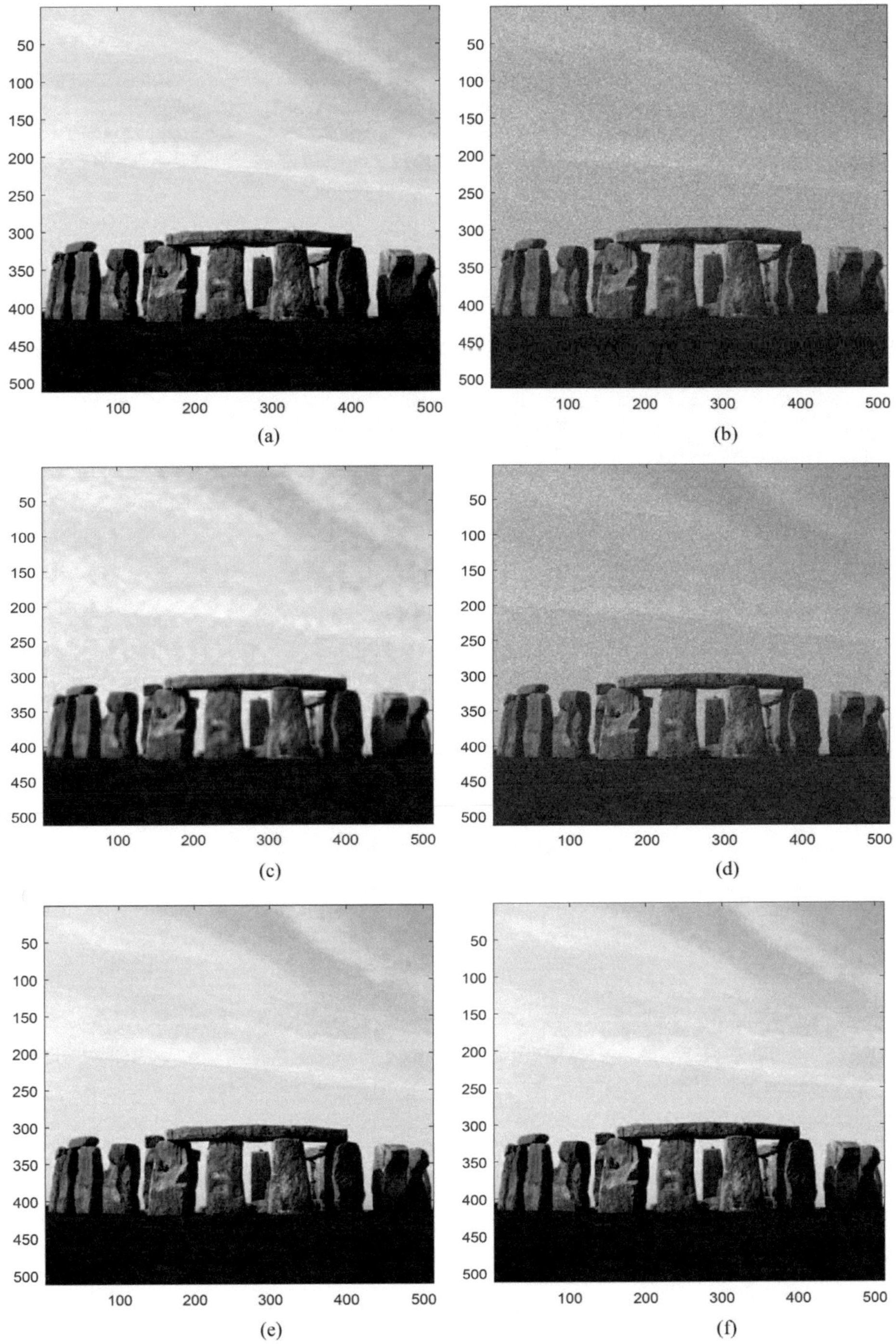

Figure 6.
An example of image denoising: (a) clean image, (b) Noisy image ($\sigma = 20$), (c) Denoised image obtained by applying the proposed technique (PSNR = 30.6924), (d) Denoised image obtained by applying deep learning *(PSNR = 22.4978), (e) Denoised image obtained by applying soft thresholding in the SWT domain (PSNR = 30.1884) (f) Denoised image obtained by applying soft thresholding in the domain of 2-D dual-tree discrete wavelet transform (PSNR = 30.2840).*

considerably reduced while preserving the original image and this precisely when applying the image denoising approach proposed in this work.

According to **Table 1**, the best results are highlighted in purple color and they are practically obtained by applying the proposed image denoising approach. Consequently, this proposed approach outperforms the other three techniques [4, 5, 9], used for our evaluation.

6. Conclusion

In this chapter, we proposed a new image denoising approach. It consists in applying a Stationary Wavelet Transform (*SWT*) based image denoising technique [4] in the domain of 2–*D* Dual-Tree Discrete Wavelet Transform. In fact, this proposed approach consists firstly in applying the 2–*D* Dual-Tree Discrete Wavelet Transform to the noisy image and the obtained noisy wavelet coefficients are then denoised by applying to each of them a *SWT* based image denoising technique [4]. Finally, the denoised image is reconstructed by applying the inverse of the 2–*D* Dual-Tree Discrete Wavelet Transform to the obtained denoised wavelet coefficients. The performance of this proposed image denoising approach, is proved by the results obtained from the computations of PSNR (Peak Signal-to-Noise Ratio) and SSIM (Structural Similarity). In fact, it permits to obtain the best values of PSNR and SSIM compared to three other image denoising techniques existing in literature. These three techniques are as follows: the Image denoising approach based on neural network [9], the denoising technique based on thresholding in the *SWT* domain [4], and the denoising approach based on thresholding in the domain of 2-DDual-Tree Discrete Wavelet Transform [5]. For example, when the noisy image is Noisy House.tif with standard deviation of Additive Gaussian White Noisen, ($\sigma = 20$), the proposed denoising approach permits to obtain $PSNR = \mathbf{30.7882}$ and $SSIM = \mathbf{0.8462}$. However, the denoising technique based on thresholding in the SWT domain [4], permits to obtain $SNR = 30.3082$ and $SSIM = 0.7360$. The denoising technique based on deep learning [9], permits to obtain $PSNR = 22.4921$ and $SSIM = 0.3090$. The denoising technique based on thresholding in the domain of 2-D Dual-Tree Discrete Wavelet Transform [5], permits to obtain $PSNR = 30.5116$ and $SSIM = 0.7766$. Consequently, this example shows clearly the superiority of the proposed denoising approach compared to the other three techniques existing in literature [4, 5, 9]. The main drawbak of this proposed image denoising approach is its computation complexity compared to the two other denoising techniques [4, 5]. This complexity is due to the fact that we apply the *SWT* based image denoising technique [4] not to the whole noisy image but we apply this technique [4] to each noisy coefficient obtained from the application of the 2–D Dual-Tree Discrete Wavelet Transform, to the noisy image. Moreover, in this proposed denoising approach, we use two completely different wavelet transforms which are the Stationary Wavelet Transform (SWT) and the 2–D Dual-Tree Discrete Wavelet Transform. Consequently, we have more computation complexity compared to the two other image denoising techniques [4, 5].

Author details

Mourad Talbi[1]* and Riadh Baazaoui[2]

1 LaNSER, Center of Researches and Technologies of Energy of BorjCedria, Tunis, Tunisia

2 Faculty of Sciences of Tunis, Department of Mathematics, Tunisia

*Address all correspondence to: mouradtalbi196@yahoo.fr

References

[1] Tian C, Fei L, Zheng W, Xu Y, Zuo W, Lin CW. Deep learning on image denoising: An overview. Neural Networks. 2020a;**131**:251-275

[2] Ismael AA, Baykara M. Digital image Denoising techniques based on multi-resolution wavelet domain with spatial filters: A review. Traitement du Signal. 2021;**38**(3): 639-651. Available from: http://iieta.org/journals/ts

[3] Mittal N, Raj A, Goyal H. Enhancement and removal of noise from medical images by wavelet transform method. In: 2019 3rd International Conference on Electronics, Communication and Aerospace Technology (ICECA). Coimbatore, India: IEEE Xplore; 2019. pp. 1126-1130. DOI: 10.1109/ICECA.2019.8821979

[4] Generate MATLAB Code for 2-D Sta tionary Wavelet Denoising - MATLAB & Simulink - MathWorks France. Available from: https://www.mathworks.com/help/wavelet/ug/generating-matlab-code-for-2-d-stationary-wavelet-denoising.html

[5] Wavelet Software at Brooklyn Poly (nyu.edu). Available from: https://eeweb.engineering.nyu.edu/iselesni/WaveletSoftware/

[6] Patrick L. Ondelettes complexes pour l'analyse des lois d'echelles. Master des Sciences de la Matière, Ecole Normale Supérieure de Lyon Université Claude Bernard Lyon 1; 2006

[7] Al Jumah A. Denoising of an image using discrete stationary wavelet transform and various thresholding techniques. Journal of Signal and Information Processing. 2013;**4**:33-41. DOI: 10.4236/jsip.2013.41004

[8] Demirel H, Anbarjafari G. Image resolution enhancement by using discrete and stationary wavelet decomposition. IEEE Transactions on Geoscience and Remote Sensing. 2011; **20**(5):1458-1460

[9] Denoise image using deep neural network - MATLAB denoiseImage (math works.com). Available from: https://www.mathworks.com/help/images/ref/denoiseimage.html

[10] Talbi M, Bouhlel MS. Wavelets and LPG-PCA for image denoising, wavelet theory and its applications. 2018. ISBN: 978-1-78923-433-6, DOI: 10.5772/intechopen.7445